MALARIA

MALARIA

A PUBLICATION OF THE TROPICAL PROGRAMME OF THE WELLCOME TRUST

Edited by

A. J. KNELL

Oxford   New York   Tokyo
OXFORD UNIVERSITY PRESS
1991

Oxford University Press, Walton Street, Oxford OX2 6DP
Oxford New York Toronto
Delhi Bombay Calcutta Madras Karachi
Petaling Jaya Singapore Hong Kong Tokyo
Nairobi Dar es Salaam Cape Town
Melbourne Auckland
and associated companies in
Berlin Ibadan

Oxford is a trade mark of Oxford University Press

Published in the United States
by Oxford University Press, New York

British Library Cataloguing in Publication Data
Malaria.
1. Man. Malaria
I. Knell, A. J.   II. Wellcome Tropical
Institute
616.9'362
ISBN 0–19–854742–0

Library of Congress Cataloging in Publication Data
Malaria: a manual from the Wellcome Tropical Institute/edited by
A. J. Knell.
'Derived from the Malaria Exhibition compiled and created in the
Wellcome Tropical Institute, and displayed there in 1987 and 1988'
—Pref.
Bibliography.   Includes index.
1. Malaria—Popular works.   2. Malaria—Atlases.   I. Knell, A. J.
II. Malaria Exhibition (1987–1988: Wellcome Tropical Institute)
III. Wellcome Tropical Institute.
[DNLM: 1. Malaria—atlases.   2. Malaria—exhibitions.   WC 17 M237]
RC156.M375   1989   614.5'32—dc19   89–2848
ISBN 0–19–854742–0 (pbk.)

Typeset by Oxford Text System
Printed in Hong Kong

Preface

This book is derived from the Malaria Exhibition compiled and created in The Wellcome Tropical Institute, and displayed there in 1987 and 1988. The exhibition was intended to bring together much new information on malaria, some of it for the first time, and to increase awareness of the global importance of this resurgent disease. Reworking the material into a book has allowed revision where necessary.

Malaria is an enormous subject, touching many different specialties, from electron microscopy to civil engineering and sociology. It is not possible to give all topics equal coverage, and I have chosen a biological emphasis. The specialized clinical problem of the management of severe *falciparum* malaria is included as an appendix. The book ends with a list of reference books and a selection of recent reviews, which provide more comprehensive information.

Many people contributed time, knowledge, experience or material to the exhibition and to this book. I acknowledge with gratitude their generosity and invaluable assistance.

I wish to acknowledge in particular the new and important material contributed by the following. Dr J. F. G. M. Meis, Professor R. E. Sinden, and Dr L. H. Bannister for the electron micrographs illustrating the complete life cycle for the first time in this form; Dr J. D. Charlwood and Dr H. Townson for their material from Papua New Guinea; Professor D. Bradley for material in the chapter on epidemiology; Professor G. Davidson for the section on insecticides; and Dr Zhao-Xi Zhou for the section on malaria eradication in the People's Republic of China. During the production of exhibition and book Professor Leonard Bruce-Chwatt has been my 'essential malariologist', Professor Eldryd Parry my enthusiastic supporter, and Peter Cheese my constant source of technical advice and ideas.

Most original charts and diagrams in this book were created by Jennie Smith, designer at the Wellcome Tropical Institute. The strength of her work made feasible the transformation of the exhibition into a book. The original illustrations in the section on insecticides and malaria are the work of Andrew Koske of Nairobi. Sheila Aspinall managed and organized the materials with unfailing efficiency and Sue Bramley organized the copyright clearance for each item in the huge collection.

The Wellcome Tropical Institute is financed solely by the Wellcome Trust, which provided further financial support for production of the book.

A. J. KNELL
Wellcome Tropical Institute

In 1730 Dr Thomas Fuller wrote:

'Can any man, can all the Men in the World, tho' assisted by Anatomy, Chymistry, and the best Glasses, pretend positively and certainly to tell us, what particles, how sized, figured, situated, mixed, moved, and how many of them, are requisite to produce a quartan ague, and how they specifically differ from those of a tertian . . . ?'

We are now able to tell all these things. They have been written in hundreds of books, and are familiar to thousands of students. Those who belittle the powers of science are not always, perhaps, the wisest of men.

The history of malaria contains a great lesson for humanity—that we should all be more scientific in our habits of thought, and more practical in our habits of government.

SIR RONALD ROSS 1910

Contents

1   Introduction

1.1   What is malaria?

Malaria is an infectious disease transmitted by mosquitoes. It is caused by minute parasitic protozoa of the genus *Plasmodium*, which infect human and insect hosts alternately.

Figure 1 is a plan of the life cycle of *Plasmodium*. Only female mosquitoes suck blood, and parasites infect a human while a mosquito of the genus *Anopheles* is feeding (1). In humans the parasites multiply dramatically, first in the liver (2), then in the blood (3). Other *Anopheles* females are infected if they suck blood from an infected human (4). The parasites multiply again in the stomach-wall of the

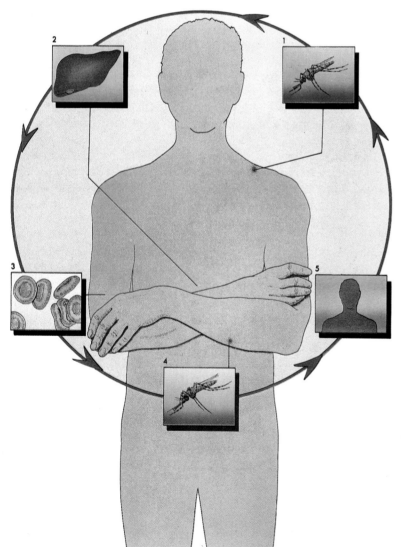

1. The breeding cycle of the malaria parasite *Plasmodium* in man and mosquito.

1. An infected female *Anopheles* bites, injecting *Plasmodium* parasites into the blood. They pass quickly into the liver.
2. The parasite multiplies in liver cells over the next 7–10 days, causing no symptoms.
3. Parasites burst from the liver cells to invade erythrocytes and multiply again. The new parasites invade more erythrocytes. This cycle is repeated, causing fever each time parasites break free and invade.
4. If a female *Anopheles* feeds on this patient, parasites will multiply in her stomach wall. Thousands of new parasites migrate to her salivary glands, to be injected in saliva when next she feeds.
5. The mosquito inoculates another human.

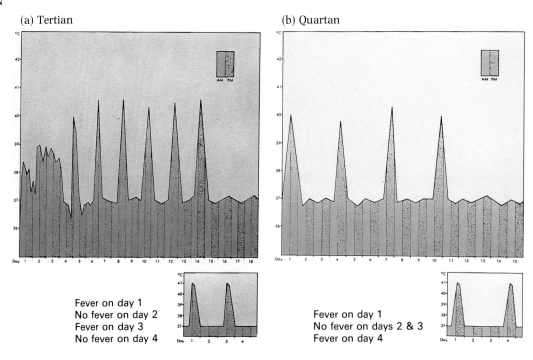

(a) Tertian   (b) Quartan

Fever on day 1
No fever on day 2
Fever on day 3
No fever on day 4

Fever on day 1
No fever on days 2 & 3
Fever on day 4

2. Tertian and quartan fevers.

(a) Fever chart from a case of *Plasmodium vivax* malaria. After an initial irregular fever the synchronized development cycles of the parasite produce paroxysms of fever on alternate days. This illness was terminated by quinine.

(b) Quartan fevers in *Plasmodium malariae* infection.

Trustees of the Wellcome Trust

3. The three stages of an attack of malaria.

Cartoons by a British serviceman in North Africa during the war.

Royal Air Force Institute of Pathology & Tropical Medicine, Halton

mosquito, then migrate through her body to infest the salivary glands. When the mosquito feeds again she injects saliva containing parasites into another human (5), and the cycle starts again.

The typical symptom of malaria is a violent fever lasting 6–8 hours, recurring every two or three days. The different species of *Plasmodium* cause two types of intermittent fever (Fig. 2). A tertian fever has one day free of fever between paroxysms; a quartan fever has two. Anaemia and enlargement of the spleen develop as the disease progresses.

These periodic paroxysms of prostrating fever distinguish malaria from other infections. Each attack is abrupt and often severe. The patient feels very ill, with headache and backache. A feeling of unbearable cold comes on rapidly, and causes violent, uncontrollable shivering. Within an hour or so the body temperature rises to 40–41°C (105–106°F). A feeling of unbearable heat follows, and lasts for another 1–2 hours. Profuse sweating then ends the attack, restoring the body temperature some 5–8 hours after it began to rise. Each attack exhausts the victim, but between attacks there are few other symptoms. Figure 3 shows the three stages of an attack of malaria, drawn by a serviceman in North Africa during the war.

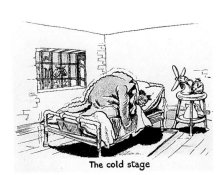

The cold stage

The hot stage

The sweating stage

The lethal form of malaria, now called severe *falciparum* malaria, presents an irregular tertian fever; but the clinical picture is dominated by rapid deterioration of the patient into stupor, fits, and coma (Fig. 4), usually leading to death.

Malaria is the world's worst health problem. At this moment more people are ill with malaria than any other disease, and the numbers affected or at risk are increasing remorselessly. Hopes that malaria might be eradicated proved impossible to realize. Control measures are becoming less effective, and the threat of epidemic malaria is increasing in many tropical areas.

Each year severe *falciparum* malaria causes up to 2 million deaths, most of them in children. Now it is largely a tropical disease, but until 25 years ago summer epidemics occurred in many temperate areas, including the Mediterranean countries and Holland. Malaria is a recurring theme in European art and literature. Daumier's drawing (Fig. 5) shows the debilitation caused by chronic malaria. Before this century malaria occurred in the South of England in summer and early autumn. The intermittent fevers were called 'agues', a word used several times by Shakespeare.

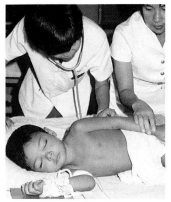

4. A Thai child dying in coma caused by severe *falciparum* malaria.   *Prof. D. A. Warrell*

1.2 Malaria in human history

Malaria is probably older than man, so human *Plasmodium* species must have evolved with man in Africa. *Falciparum* malaria may be an exception, adapting itself to man relatively recently. Malaria spread with human migration in the neolithic period into Europe, the Middle East, Asia, India and China (Fig. 6). Its spread to Central and South America was probably from Asia in pre-Columbian times, perhaps in the first millennium AD. It is uncertain if malaria reached the Caribbean islands from the mainland, or if it was introduced by the slave trade.

5. 'The Sick Peasant' by the nineteeth-century French artist Daumier.   *Éditions André Sauret*

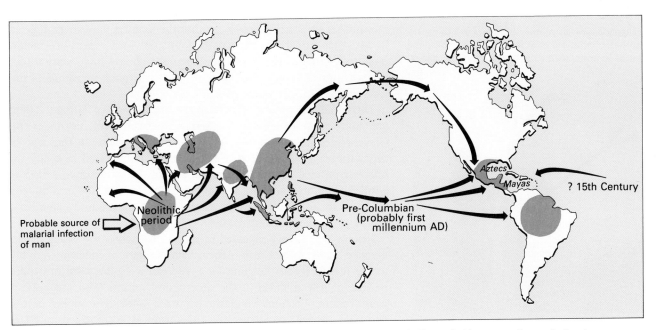

6. The probable routes of spread of malaria.
Trustees of the Wellcome Trust

तुङ्गीनासो विचिलकद्दालको वाहकस्तथा ।
कोष्ठागारी क्रिमिकरो यत्र मण्डलपुच्छक: ॥
तुङ्गनाम: सर्पपिकोऽवल्गुली ग्रन्थुकस्तथा ।
च्यमिकीटश्च घोरा: स्युर्दंदश प्राणनाग्रना: ॥
तैमंबन्लीह दृष्टानां वेगब्नानानि सर्पवत् ।
तास्त्राश्च वेदनास्तीव्रारोगा वै सान्निपातिका: ॥
चाराग्निदग्धवद्ग्रे रक्तपीतसिताश्या: ।
ज्वराङ्गमर्दरोमाश्ववेदनामि: समन्वित: ।
छर्द्युत्तीसारतृष्णा च दाहो मोहविजृभिका ॥
वेपथुश्वासहिक्काश्च दाह: शीतश्च दारुणम् ।
पिडकोपचय: ग्रोफो यन्यथो मण्डलानि च ।

मश्यका: सामुद्रा: परिमण्डलो हस्तिमश्यक: कृष्ण: पाइर्श्वतीय इति
पश्च । निर्दिष्टश्च तीव्रकएदूरंग्रोफश्च । पार्श्वतीयत्त कीट: प्राणहरैसत्म-
लचग: ॥

7. A Sanskrit passage from Susruta (5th century BC). It describes twelve lethal forces, including acute fevers, and the mosquitoes which cause them.

Athlone Press

8. Hippocrates (5th century BC) described the association of intermittent fevers with marshes, and clinical features, such as enlargement of the spleen.

Trustees of the Wellcome Trust

9. Men of the 231 brigade antimalaria unit spraying mosquito breeding sites in Sicily in 1943. The spray contained arsenic and protective clothing was essential.

Imperial War Museum

The wide distribution of malaria is attributable to the adaptability and breeding potential of the *Anopheles* mosquito. By contrast, dependence on the much more fastidious tsetse fly has restricted *Trypanosoma brucei* and sleeping sickness to riverine and rain-forest areas of Africa.

Indigenous populations in Europe and other temperate zones did not develop constitutional resistance to malaria, because it remained a disease of summer epidemics. Malaria is recorded in ancient writings from Asia and China (Fig. 7). Throughout history malaria has plagued travellers to the tropics—merchants, missionaries, and soldiers. It has affected the course of history many times.

An early example is from the classical Greek historian Thucydides. He tells how in 413 BC the best part of the Athenian army besieged the Sicilian port of Syracuse, an ally of Athens' main enemy, Sparta. Initial failures dashed hopes of early victory and the Athenians were forced in late summer to occupy marshy ground around Syracuse. Reports from Syracuse told the Athenians that factions in the city were plotting a coup, and if successful they planned to surrender to Athens. An eclipse of the moon was considered an ill omen, giving further reason for delay.

The Athenians then suffered a lethal epidemic. Soldiers became severely ill with violent fever, and many died in a few days. When the delayed attack on Syracuse was ordered, it was too late. An Athenian naval assault was defeated, and the army was so reduced and weakened by disease that an attempt to break off and escape led to its destruction by the forces of Syracuse. The few survivors were enslaved, and none escaped to Athens.

Athens never recovered from this catastrophe, and the inevitable defeat by Sparta was not long delayed. Her overthrow led to the collapse of Classical Greece, and eventually permitted Rome to become the dominant Mediterranean power.

The epidemic was almost certainly malaria and the spies' story a deception. The Syracusans must have known the risks of camping in the marshes in late summer. They probably did not know that mosquitoes spread the disease, or that the malaria parasite completes its life cycle rapidly in hot weather, but previous experience would have taught them the risk. The Athenians should have known too: the writings of Hippocrates (Fig. 8) contain explicit warnings of the risks of fever in wet places in summer.

Thus malaria influenced decisively the history of Europe. There are many more examples. The remarkable career of Alexander the Great ended abruptly in 323 BC, when he died at the age of thirty-three, probably of cerebral malaria. Rome itself is a city on hills surrounded by the former marshes of the Campagna, a trap of pestilence for attackers on several occasions. Indeed 'Mal'aria' means 'bad air' in Italian. It arose from the belief that malaria came from putrid air above stagnant water.

History was nearly repeated in July 1943, when the allied armies invaded Sicily from North Africa. The dangers of malaria, notably to the British troops heading for Syracuse, were foreseen, but there had been little malaria in the desert, and antimalarial precautions were casual. Epidemic malaria occurred after two weeks of the five-week campaign. Medical facilities were severely tested by 6361 cases, each staying 3 weeks in hospital, on average. Only 13 died, and anti-malarial units soon controlled the disease (Fig. 9), but British battle and malaria casualties were roughly equal.

The British avoided disaster; the Athenians did not. Knowing one fact made the difference: **mosquitoes transmit malaria.**

1.3   The history of malaria

Peruvian bark and quinine

The first important event in the history of malaria was the discovery of the 'Peruvian fever tree', *Cinchona* (Fig. 10). In the early 17th century Jesuit missionaries in South America learned of the medicinal value of *Cinchona* bark to treat and cure fevers, probably from Inca herbalists. Its fame spread rapidly, and a legend persists that the Countess of Chinchón, wife of the Viceroy of Peru, was cured of tertian fever by drinking an infusion of the bark. In about 1637 Cardinal Juan de Lugo and other Spanish priests brought the new remedy to Rome, and used it in the Hospital Santo Spirito (Fig. 11). Soon the bark preparation, known as 'Jesuits' powder' became widely used in Europe. The British and Dutch introduced it to India, and in 1692 missionaries used it to cure the Chinese Emperor of a malignant fever.

By 1677 the London Pharmacopeia had an entry for *Cortex Peruvianus* (Fig. 12). In 1666 Morton and Sydenham recognized the specific action of the bark in 'agues' (intermittent fevers), observations confirmed and clearly described in 1712 by Torti in Italy. However, many physicians continued to prescribe the bark indiscriminately, until James Lind described in 1765 the most effective method of its use.

Linnaeus named the tree producing Peruvian bark *Cinchona* in 1749. In 1820 Pelletier and Caventou in Paris isolated the two main alkaloids, quinine (Fig. 13) and cinchonine. This allowed dosage to be accurately prescribed, and the assay of the alkaloid content of different batches of bark.

Demand for *Cinchona* bark rapidly exceeded supply. In 1850 the British government in India needed some 9 tons of quinine annually. Yet all bark had to be carried by men or pack animals over hundreds of miles of Andean terrain to the seaports of South America. The reckless cutting of wild *Cinchona* trees exhausted the natural resource. Attempts to establish *Cinchona* plantations in other parts of the world failed, not least because the collection and export of good *Cinchona* seeds was opposed by governments and indigenous people.

The solution of the production problem began in 1861. Charles Ledger (Fig. 14), a trader and entrepreneur in Peru, sent his servant, Manuel Incra Macrami (Fig. 15), to collect seeds from an exceptional stand of *Cinchona* trees, which they had discovered on the eastern slopes of the Bolivian Andes some years earlier. For three years Manuel camped in the forest, seeing the flowers and seeds of the trees destroyed by late frosts each season. In the fourth year his patience was rewarded by a heavy crop. He packed two leather bags with seed from about fifty of the best trees, trecked several hundred miles over the Andes to Ledger's home in Tacna, and delivered the bags to Ledger during the night of 19 May 1865.

Ledger sent about 15 kg of seeds to his brother George in London. Attempts to interest the British authorities failed, but the Dutch consul bought one pound of seed for 100 guilders and a promise of a further sum if the germination proved successful. The seeds were

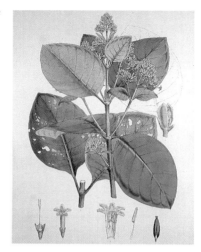

10. Leaves and flowers of *Cinchona succirubra*, an original species of 'fever-bark' tree.

11. Fresco from the Santo Spirito Hospital. The inscription says 'The purple-clad father comforts the sick with the fever-bark of Lima'.   *Greenberg Publishers*

88   TINCTURÆ.

TINCTURA CINNAMOMI.

℞ Cinnamomi ꝑ. fefcunciam,
    Spiritûs vinofi tenuioris ℳ. libram unam.

    DIGERE fine calore, et cola.

TINCTURA CORTICIS PERUVIANI
simplex.

℞ Corticis Peruviani ꝑ. uncias quatuor,
    Spiritûs vinofi tenuioris ℳ. libras duas.

    DIGERE, et cola.

TINCTURA CORTICIS PERUVIANI
volatilis.

℞ Corticis Peruviani ꝑ. uncias quatuor,
    Spiritûs falis ammoniaci ℳ. libras duas.

    DIGERE fine calore in vafe bene claufo, et cola.

TINCTURA FOETIDA.

℞ Afæ fœtidæ ꝑ. uncias quatuor,
    Spiritûs vinofi rectificati ℳ. libras duas.

    DIGERE, et cola.

                                    TINCTURA

12. A page from the London Pharmacopoeia of 1677.

- ◯ Nitrogen
- ● Carbon
- ◉ Oxygen
- ◯ Chlorine

13. The structure of quinine, the first chemotherapeutic compound, and of three modern antimalarial drugs.

Quinine, the original antimalarial drug, and still one of the best. Quinine is one of the active principles in the bark of *Cinchona ledgeriana*, a tree native to the eastern foothills of the Andes.

Quinine is a quinoline derivative, the structure basic to a large group of antimalarial compounds.

Chloroquine, a typical antimalarial drug of the 4-aminoquinoline type.

First discovered in Germany in the 1930s, chloroquine played a vital part in the allied war effort in the Far East. For many years chloroquine was one of the safest and most widely used antimalarials. Now its value is seriously restricted by resistance.

Proguanil was produced in the British programme to find new antimalarials during the Second World War. It is a very useful drug, but its value is limited by resistance.

Proguanil is converted to the active compound cycloguanil by metabolism in the body. It is the most important anti-folate antimalarial.

Trustees of the Wellcome Trust

sent to the Dutch plantations in Java, arriving in December 1865. Some 20 000 germinated, and George Ledger promptly received an additional 500 guilders, as promised. About 12 000 trees were reared, and in 1872 were sufficiently mature for assay of the bark. Values of 4 to 8 per cent quinine meant that the new trees produced up to ten times more than the older varieties in cultivation. Two years later the bigger trees were producing up to 13 per cent quinine. Wisely the Dutch destroyed their existing plantations and by selective breeding increased the yield from the new variety. Production of quinine grew rapidly, and its price fell by 90 per cent over the next 20 years, giving the Dutch a profitable monopoly.

The remainder of Ledger's seeds were wasted. In 1871 Manuel went back to Bolivia in search of more seed for Ledger. On his way back the police arrested him. Manuel was ill-treated but refused to reveal Ledger's name. After 20 days he was released, but died a few days later. Ledger received little compensation for his enterprise, and was eventually to die in poverty in Australia; but the new species was named *Cinchona ledgeriana* in his honour.

The malaria parasite and mosquitoes

The vital discovery that mosquitoes transmit malaria was made by Ronald Ross on the 20 August 1897, in Secunderabad, in India. Ross was then 40, a Surgeon Major in the Indian Army (Fig. 16). He had worked for two years to prove that mosquitoes transmit malaria, as suggested to him by Sir Patrick Manson. Manson was the leading specialist in tropical diseases in London (Fig. 17), and had done much to prove that mosquitoes spread filariasis, a nematode worm disease. The idea that mosquitoes transmit malaria had been suggested before, notably by the Roman physician Lancisi in 1717, and by Albert King in the USA in 1882, before Manson developed the theory in detail in 1894.

Manson showed Ross microscopic preparations of the tiny parasites in the red blood cells of malaria patients, discovered in 1880 by Alphonse Laveran (Fig. 18), a young doctor in the French Army in Constantine, in Algeria. Necropsy studies of patients dying of malaria had revealed black pigment in blood cells and organs, containing

14. Charles Ledger. *Royal Pharmaceutical Society of Great Britain*

16. Ronald Ross at the Nobel Prize ceremony in 1902. *Ross Institute*

17. Sir Patrick Manson. *Trustees of the Wellcome Trust*

15. Manuel Incra Macrami. *Royal Pharmaceutical Society of Great Britain*

18. The Laveran Medal of the Société de Pathologie Exotique in Paris.

20. Drawings by Ross of oocysts on the stomach wall of a *Culex* mosquito.

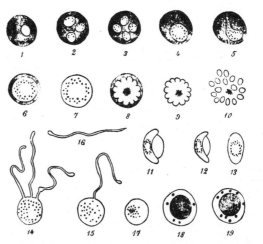

19. Laveran's first drawings of malaria parasites in blood from a soldier with acute fever.

iron derived from the digestion of haemoglobin by the parasite. Laveran was convinced that malaria is a disease of red blood cells, but his microscope magnified only 400 diameters, and the malaria parasites are only just visible at this power. His discovery was a triumph of patient observation, leading to success where many others had failed (Fig. 19). He had great difficulty in persuading the medical authorities that these tiny objects were indeed parasites, and that the mysterious cause of malaria had been found. Eventually Pasteur journeyed to visit Laveran, and was deeply moved by his demonstration. Even so, in 1897 many were still sceptical, including Ross himself, until convinced by Manson.

Back in India, Ross searched for malaria in mosquitoes fed on patients. He persisted through two years of disappointment and frustration, encouraged by regular letters from Manson. He soon realized that different species of mosquito abound, and only one might transmit malaria, but he lacked any means of systematic identification. Then on 15 August 1897 his servant brought some unusual mosquito larvae: smaller than most and lying parallel to the surface of the water in which they lived. Next day these larvae had become brown mosquitoes with dappled wings, resting with heads down and back legs raised. Ross fed ten of them on his patient, Hussein Khan. By 20 August only two were still alive. He dissected one, and with his microscope saw something new: clear cysts on the outer wall of the stomach (Fig. 20). On closer examination he saw that the cysts contained feathery black pigment, which he recognized. It was identical to the pigment in human erythrocytes parasitized by malaria.

The next day he dissected the second: he saw the cysts again, now larger, and again containing pigment. He had found malaria in mosquitoes. We now know that the reason it had taken so long is that naturally he had concentrated on the large grey mosquitoes which are easily seen by day. These are *Culex* and related genera, which do not transmit malaria. The small brown *Anopheles* is secretive and flies at night.

Ross wrote Manson an excited letter describing his new findings, including sketches. Manson reported Ross's discovery to a meeting of the British Medical Association in Edinburgh. At this point Ross was posted by the Indian Army to an area free of malaria, but after several months of vigorous protest he was allowed to go to Calcutta to continue his researches. Here too human malaria was scarce, but bird malaria was common. Using this convenient model Ross rapidly

21. Major and Mrs Ross, his assistant Mohammed Bex, and another, in Calcutta in 1898. The cages at the foot of the steps are for the birds used by Ross to establish the transmission of bird malaria by mosquitoes.   *Trustees of the Wellcome Trust*

established the transmission of malaria during the feeding of infected mosquitoes (Fig. 21). In 1902 he received the Nobel Prize.

In the meantime the discovery was confirmed by Grassi (Fig. 22) and his team in Rome. Grassi identified the malaria mosquito as *Anopheles*, demonstrated insect transmission from man to man, and showed that protection from *Anopheles* protects from malaria. With Manson, Grassi's colleagues Bignami and Bastionelli did an important experiment. They sent to London mosquitoes infected by feeding on *vivax* malaria patients in Rome. Manson allowed them to feed on volunteers (his son, Thorburn Manson, and George Warren, a laboratory assistant). Both developed malaria. The vector theory of disease transmission was established.

It was a revolution in medical science comparable to the discovery of the Double Helix in 1953. It created a new specialty of Tropical Health and Hygiene. Institutions were founded to develop and teach the new ideas. Pilot schemes to control malaria by attacking mosquitoes succeeded. One of the earliest began in Khartoum in 1902, when Henry Wellcome sent Andrew Balfour to do research and to improve health in the Sudan. By applying the new knowledge the United States was able to build the Panama Canal, after the French abandoned the project in the face of appalling human losses from malaria and yellow fever, a virus causing hepatitis, also spread by mosquitoes.

This book tells how our understanding of malaria has increased since those early discoveries. There are electron micrographs of every stage in its complicated life cycle. Mosquito biology is complicated but fascinating, and still poorly understood. The dash for global eradication of malaria in the 1960s and 70s made good initial progress but eventually foundered. We know that new drugs and insecticides developed at great cost tend to lose their value as the parasite or the mosquito becomes resistant. Molecular biology now offers hope of malaria vaccines, but few believe these will be decisive in the battle against malaria, still the world's worst health problem.

22. Giovanni Battista Grassi (1854–1925).

2 Malaria: the problem

2.1 Four species of malaria parasite infect humans

Four species of *Plasmodium* infect man. Three may cause severe illness, but are rarely fatal. These are the types of malaria which cause distinctive intermittent fevers, and a patient's temperature chart is a reliable guide to diagnosis.

- *Plasmodium vivax* (tertian).
- *Plasmodium ovale* (tertian).
- *Plasmodium malariae* (quartan).

The fourth causes much more serious and progressive illness, often leading to coma and death within a few days (Fig. 23).

- *Plasmodium falciparum.*

This has had many names, including aestivo-autumnal fever, malignant tertian malaria (MT), and cerebral malaria. *Falciparum* malaria is the name now preferred. Blackwater fever is an unusual variant, which typically occurred in white residents in malarious areas, who had taken quinine erratically to prevent malaria over long periods. If such a person caught *falciparum* malaria, and especially if quinine was given, then rapid destruction of erythrocytes (haemolysis) caused haemoglobin release into the urine.

Plasmodium falciparum is the most common malaria parasite in Africa. It is especially dangerous to children aged between one and five, and to travellers from temperate areas. Indigenous adults are protected by a form of partial immunity called premunition. This develops slowly, and only in response to repeated infections by a specific species of *Plasmodium*. It is lost if continuous exposure to infection is not maintained. In such individuals parasites are commonly found in the blood without symptoms appearing.

Thus malaria causes three clinical patterns, according to the species of infecting parasite and the previous exposure of the individual to malaria.

- *Intermittent fevers* or *severe falciparum malaria* (*cerebral malaria*) occur especially in non-immune people, such as travellers and inhabitants of areas where epidemics of malaria break out from time to time.
- *Asymptomatic parasitaemia* is common in people indigenous to endemic areas, who have developed premunition.

In clinical diagnosis the types of malaria are usually distinguished by their species names, as in *vivax* malaria, or *falciparum* malaria.

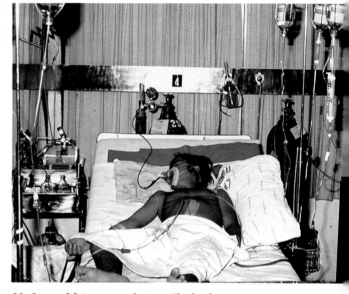

23. Severe *falciparum* malaria in Thailand.
Intensive care failed to save this Thai child in coma with severe *falciparum* malaria.   *Prof. D. A. Warrell*

2.2   Malaria is a difficult disease to study

For much of the life cycle malaria parasites are very small. One stage is smaller than most bacteria, another grows within erythrocytes, which are only 7.2 microns in their greatest diameter. The life cycle is complicated. Laboratory culture was not achieved until 1976, for *Plasmodium falciparum* only, and is difficult and expensive. The liver stages are inaccessible: in principle infection can be established by just one parasite in an organ weighing 1.8 kg. The blood forms need special stains to be seen clearly under the microscope.

The female *Anopheles* mosquito is inconspicuous, smaller than many other mosquito genera, and secretive in its habits. It feeds and is active at night, avoiding light. Its bite mark is relatively small. Precise identification of *Anopheles* species may require complicated laboratory techniques. An effective vector species may look exactly the same as another species which rarely bites man.

Our understanding of malaria grew slowly. During the nineteenth century rapid advances were made in the identification of bacteria causing major human diseases such as typhoid fever, but the organism causing malaria remained a mystery. Then Laveran first saw parasites in the blood of malaria patients in 1880, but at first no one would believe him. Ross searched for two years before he found the mosquito phase in 1897. A liver stage had been suspected for many years, but was not demonstrated until 1948. The resting liver form responsible for delayed relapses in *Plasmodium vivax* infection was discovered in 1979. The forms responsible for delayed relapses in *Plasmodium malariae* are still unknown.

2.3   The life cycle of *Plasmodium*

The life cycle of *Plasmodium* (the malaria parasite) is shown in Fig. 24. It is easier to understand as a sequence of four phases: one sexual without multiplication, and three asexual with multiplication. *Plasmodium* has sex without reproduction, then reproduces without sex (how much simpler life would be if humans did the same). The sexual and first asexual phases occur only in *Anopheles* mosquitoes. The second asexual phase is in the liver, the third in the blood. The third phase is repeated many times. Each asexual phase begins with feeding and growth. Every phase ends when new invasive parasites appear. In the third phase some parasites become sex-cells, called gametocytes, which start a new cycle if taken into an *Anopheles*.

1. Fertilization, the sexual phase, takes place in the stomach of a mosquito.
2. Sporogony, the first asexual phase, in the stomach wall and body of the mosquito.
3. Hepatic schizogony, the second asexual phase, in the liver.
4. Erythrocytic schizogony, the third asexual phase, in erythrocytes.

Each asexual phase begins as an intracellular parasite adapted for rapid growth, called the sporocyst in the mosquito and trophozoites in humans. This feeds and enlarges, then divides into numerous invasive forms, which break free from the host cell and invade new ones to start the next phase of the cycle. The process is similar in each asexual phase, as shown in Fig. 25.

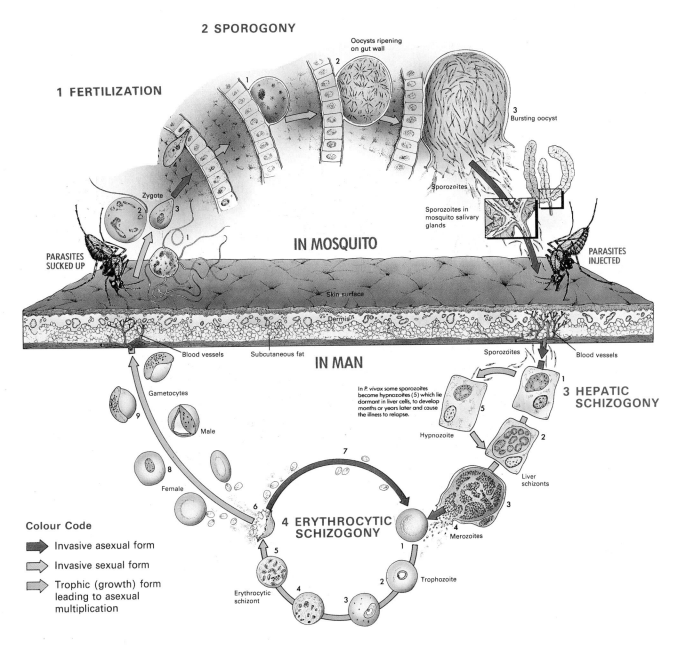

2 SPOROGONY

1 FERTILIZATION

Oocysts ripening
on gut wall

Bursting oocyst

Zygote

Sporozoites

Sporozoites in
mosquito salivary
glands

IN MOSQUITO

PARASITES
SUCKED UP

PARASITES
INJECTED

Skin surface

Dermis

Blood vessels   Subcutaneous fat   **IN MAN**   Sporozoites   Blood vessels

Gametocytes

Male

Female

In P. vivax some sporozoites
become hypnozoites (5) which lie
dormant in liver cells, to develop
months or years later and cause
the illness to relapse.

Hypnozoite

**3 HEPATIC
SCHIZOGONY**

Liver
schizonts

Merozoites

**4 ERYTHROCYTIC
SCHIZOGONY**

Colour Code

➤ Invasive asexual form

➤ Invasive sexual form

➤ Trophic (growth) form
leading to asexual
multiplication

Erythrocytic
schizont

Trophozoite

24. The life cycle of *Plasmodium*.

Phase 1. Fertilization.
A female *Anopheles* sucks a blood meal from an infected
person. Gametocytes escape from red blood cells to become
free gametes, male and female. The male gamete produces up
to 8 flagella in a sudden, violent exflagellation (1). These tear
themselves free and swim off, each with a male nucleus
attached. If a female gamete is found then fertilization (2)
produces a zygote (3). This develops into the invasive
ookinete (4), which bores into the stomach wall and becomes
an oocyst.
Phase 2. Sporogony: asexual development in the mosquito.
The oocyst grows (1), then divides to produce thousands of
invasive sporozoites (2). The mature cyst bursts (3) and the
free sporozoites migrate through the body of the mosquito and
invade her salivary glands (4).
Phase 3. Hepatic schizogony; asexual development in the
liver.
When the mosquito feeds again sporozoites are injected into

the blood. They invade liver cells (1) and become hepatic
trophozoites (2). These grow, then divide to produce
thousands of invasive merozoites (3). The infected liver cells
burst, releasing the merozoites into the blood (4). In
Plasmodium vivax some sporozoites become hypnozoites (5)
which lie dormant in liver cells, to develop months or years
later and cause the illness to relapse.
Phase 4. Erythrocytic schizogony: asexual development in the
blood.
Merozoites invade red blood cells (1) and become erythrocytic
trophozoites (2). These grow, then divide into 8–16 new
merozoites (3). When mature the red cell bursts, merozoites
are released (4), and the cycle starts again (5).
As the disease progresses some merozoites develop into male
or female gametocytes (6,7). These circulate but only develop
further if they are taken up by a mosquito.

Trustees of the Wellcome Trust

Vegetative forms
Schizogony

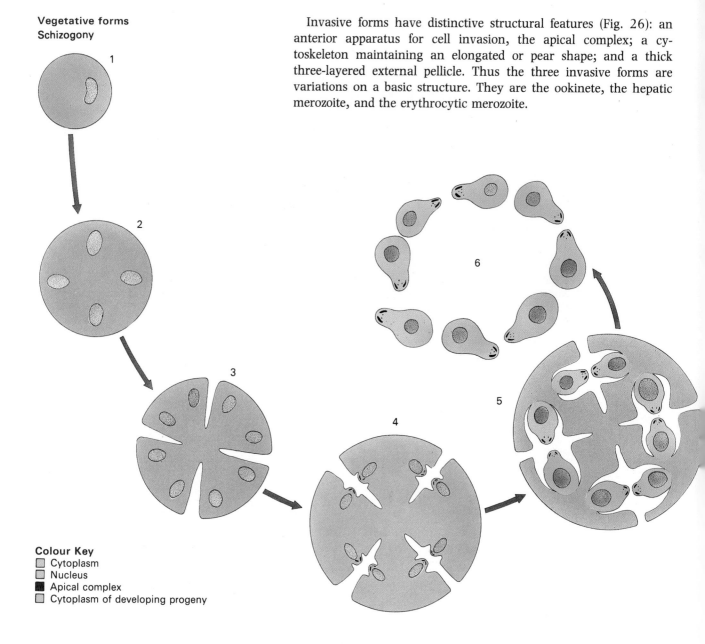

Colour Key
▢ Cytoplasm
▨ Nucleus
◼ Apical complex
▢ Cytoplasm of developing progeny

Invasive forms have distinctive structural features (Fig. 26): an anterior apparatus for cell invasion, the apical complex; a cytoskeleton maintaining an elongated or pear shape; and a thick three-layered external pellicle. Thus the three invasive forms are variations on a basic structure. They are the ookinete, the hepatic merozoite, and the erythrocytic merozoite.

25. The segmentation of asexual vegetative forms of *Plasmodium* into new invasive parasites.

Following rapid nuclear division (1,2) the cytoplasm is divided by deep clefts extending inwards from the cell membrane (3). Each nucleus becomes associated with a portion of membrane, in which the apical complex of the new invasive form develops (4). The pellicle and micro-tubules then develop (5). Finally the residual structure of the vegetative form disintegrates, releasing the new invasive parasites (6). This process is best seen in the sporocyst and hepatic trophozoite, which each produce thousands of new parasites. The erythrocytic trophozoite produces only 8–16 progeny, most of which mature in a similar way, but in association with the surface membrane itself.   *Trustees of the Wellcome Trust*

The sexual phase (Fig. 27)

Exflagellation is the explosive production of male gametes. It happens in the stomach of a mosquito within a few minutes of feeding on a malaria patient, or in fresh blood specimens (Fig. 28). Male parasites show violent internal activity, as up to 8 'flagella' form within the cell and start to beat. As the formation of flagella is completed their motion tears the cell apart, releasing them into the plasma of the blood meal, each having a nucleus attached about a third of the way from the base. The entire process takes less than a minute.

Flagella are really spermatozoa, which swim away seeking female cells. After fertilization the parasite is a zygote, which matures over the next few hours into a mobile ookinete (Fig. 29). This is an invasive form, which glides through the blood meal, and can destroy erythrocytes barring its progress. Eventually it reaches the stomach wall, where it passes between or through the epithelial cells to reach

Recurring themes in
dissimilar shapes

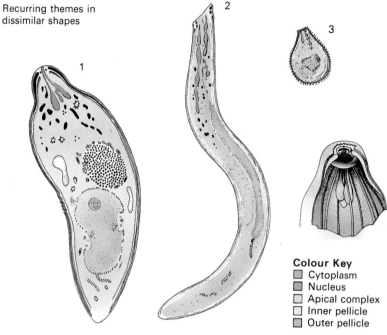

Colour Key
▢ Cytoplasm
▢ Nucleus
▢ Apical complex
▢ Inner pellicle
▢ Outer pellicle

26. Invasive forms: recurring themes in different shapes:
1. ookinete.
2. sporozoite.
3. merozoite.
The structure of the apical complex: the apparatus for invasion of a new host cell.

Trustees of the Wellcome Trust

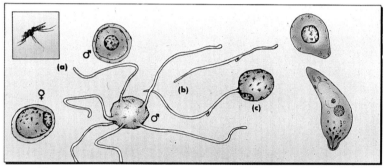

27. Fertilization.

1. Fertilization. The male and female gametes (sex cells) mature very rapidly (a) once they enter the mosquito's stomach. At the right temperature, budding and breaking away of the flagella (b) takes as little as 2 minutes. A zygote is formed (c) when a flagellum fertilizes the female gamete.

Trustees of the Wellcome Trust

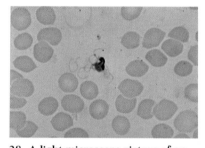

28. A light-microscope picture of an exflagellating male gamete.

World Health Organization

the basement membrane. Here it rounds up, loses the features of an invasive form, and the next phase begins.

Exflagellation, fertilization, and the ookinete are the only stages in the life cycle which can be watched, if a high-power microscope is used. Exflagellation is the event which Laveran saw, and demonstrated to visitors to Constantine, notably Pasteur. A Canadian medical student, MacCallum, first followed the flagella, saw fertilization and the ookinete, and realized that the process was sexual in nature. He first studied the parasites during a vacation in 1897, in blood from sick crows on his home farm. Later he confirmed his findings in patients with *falciparum* malaria in the Johns Hopkins Hospital. So many important discoveries about malaria were made by young people.

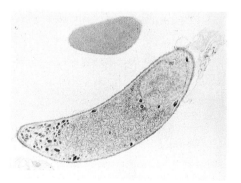

29. Electron micrograph of an ookinete.

Prof. R. E. Sinden

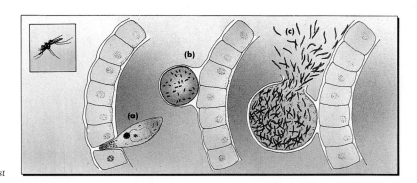

30. Multiplication in the mosquito.

In 12–24 hours the zygote becomes a leaf-shaped mobile ookinete (a). Many may penetrate to the outer surface of the stomach. Here they grow (b) and eventually burst, releasing thousands of sporozoites (c). *Trustees of the Wellcome Trust*

The first asexual phase (sporogony) (Fig. 30)

In the stomach wall the parasite grows rapidly, forming a spherical oocyst, which can be 80 microns in diameter when mature. It is enclosed by a cyst wall formed of protein secreted by the parasite and the basement membrane. The cyst projects into the body cavity (haemocoel) of the mosquito and when mature can be seen using a dissecting microscope. It feeds on the haemoglobin of the blood meal, accumulating the black malaria pigment by which Ross recognized that the cysts he had found were indeed malaria parasites. In a heavily infected mosquito the stomach may be covered in oocysts, resembling a bunch of grapes when seen under the microscope (Fig. 32).

31. Scanning electron micrograph of rupturing oocysts. Sporozoites are escaping. *Prof. R. E. Sinden*

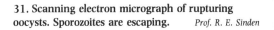

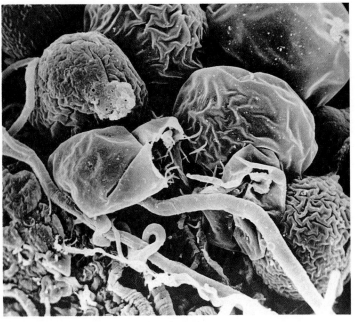

32. Scanning electron micrograph of the mid-gut of a heavily infected mosquito.

The entire organ is covered with mature oocysts.
Prof. R. E. Sinden

After about one week (depending on the ambient temperature) the oocyst begins a process of internal division, with differentiation of thousands of worm-shaped sporozoites, the next invasive form in the life cycle. The oocyst bursts fourteen or more days after infection of the mosquito (Fig. 31). The sporozoites move through the insect's haemocoel to enter her salivary glands. When the mosquito feeds again sporozoites in her saliva are injected into the blood of a new host. Within minutes they are in the liver, invading liver cells.

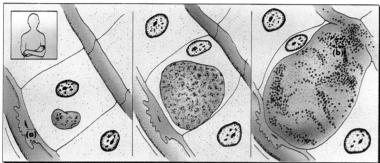

33. The liver stage (hepatic schizogony).

After injection by the mosquito, the parasites enter the liver via its Kupffer cells (a), which are usually part of the body's defence against hostile organisms and particles. Again the parasite grows and divides (b), releasing thousands of the next phase of *Plasmodium*, the merozoite.

Trustees of the Wellcome Trust

The second asexual phase (hepatic schizogony) (Fig. 33)

In a liver cell the sporozoite soon becomes the hepatic trophozoite. This absorbs nutrients from the liver cell, and at blood heat it can grow very rapidly, reaching perhaps 40 microns diameter in two days, distending and destroying its host cell as it enlarges. It now begins to divide internally, to become the multinucleated hepatic schizont (Fig. 34), in which develop many thousands of tiny, invasive merozoites. These are released when mature, about one week after the infecting mosquito bite, by rupture of the host cell into a hepatic capillary. In the blood the merozoites invade erythrocytes within a few minutes of their release (Fig. 35), and the next phase begins.

So far the infected person has no symptoms, and is unaware that infection is progressing. This is called the incubation, or pre-patent period of infection. It includes the whole of hepatic schizogony and the first few cycles of erythrocytic schizogony, the next phase.

In *Plasmodium vivax* some sporozoites entering the liver do not develop immediately. Instead they become tiny dormant parasites called hypnozoites. These persist for months or even a few years, to begin development and cause the delayed relapses typical of *vivax* malaria. *Plasmodium malariae* may cause relapses too, sometimes twenty years after, but the resting form of this species remains obscure.

The third asexual phase (erythrocytic schizogony) (Fig. 36)

In an erythrocyte the merozoite becomes the third growth form: the erythrocytic trophozoite (Fig. 37). Limited by the 7.2 micron diameter

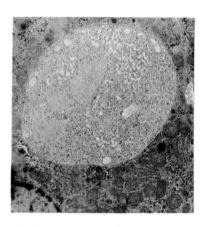

34. Electron micrograph of an early hepatic schizont.

Early schizogony in a hepatic trophozoite two or three days after sporozoite injection. Peripheral vacuolation and the formation of a labyrinth of clefts within the cytoplasm are the first signs of maturation into merozoites. *Dr. J. F. G. M. Meis*

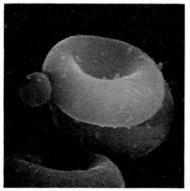

35. A merozoite caught invading an erythrocyte, seen by electron microscopy.

A merozoite (the small sphere) attached to an erythrocyte at the start of invasion, seen by scanning electron microscopy. *Dr. L. H. Bannister*

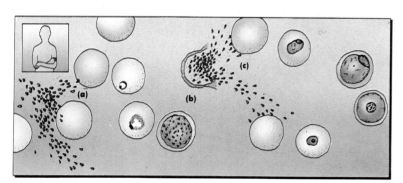

36. Breeding in the blood (erythrocytic schizogony).

The merozoites invade the erythrocytes (a). Every 2 or 3 days a new cycle of growth and division is completed (b). The schizont ruptures (c).

Trustees of the Wellcome Trust

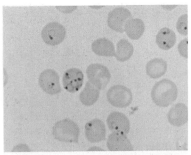

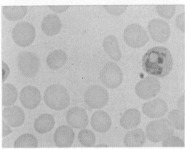

37. Erythrocytic trophozoites, as seen using light microscopy. *Plasmodium falciparum* **(top)** *Plasmodium vivax* **(bottom).**

Two pictures taken by light microscopy of blood preparations stained to show trophozoites in the erythrocytes. The top picture is *Plasmodium falciparum* and the bottom *Plasmodium vivax*.

World Health Organization

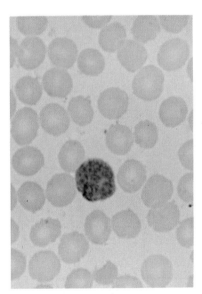

38. A fully mature schizont.

World Health Organization

of the erythrocyte, this is the smallest and least productive of the growth phases. It ingests haemoglobin and digests it, accumulating malaria pigment as it grows. After two or three days, depending on the species of *Plasmodium*, it divides to produce a rosette-shaped schizont (Fig. 38). Then the red-cell bursts, releasing 8–16 new merozoites (Fig. 39). These invade more red-cells to start a new generation in the blood.

It is these cycles of multiplication in the blood which cause the periodic fevers of malaria. As the number of parasites increases the infected person becomes ill. Fever is induced when the schizonts rupture, releasing pyrogens into the blood. The fever is irregular for a day or two, and may remain irregular in severe *falciparum* malaria. In other types the parasites soon synchronize their cycles, causing all schizonts to rupture and release merozoites at the same time. There are records of untreated *vivax* malaria which still show regular tertian fever after 25 or more cycles. Such relapsing fevers are a unique feature of malaria. They may indicate transient suppression of host immunity by the parasite at the time of merozoite release, the period most vulnerable to host defences.

In severe *falciparum* malaria the proportion of erythrocytes parasitized may rise to 30 per cent or more. In other species, or in *falciparum* malaria in an immune person, the body's defences are able to restrain the parasite, so that the number of erythrocytes parasitized is much reduced.

After several blood cycles a proportion of trophozoites develop in an alternative way, producing gametocytes (Fig. 40). These take about four days to reach maturity, but then become dormant and circulate in the blood for prolonged periods. They are the new sex-cells, which can only develop further if they are taken up by a feeding mosquito, to start a new life cycle. Gametocytes remain enclosed in the host erythrocyte membrane, breaking free only on activation in the stomach of a mosquito.

2.4 The genus *Plasmodium*, the malaria parasite

Plasmodium is a genus of parasitic protozoa in the phylum Apicomplexa, distinguished by the presence of an apical complex at some stage in the life cycle. This is an intracellular apparatus at the front end of the cell, used in the invasion of host cells. The genus *Plasmodium* is further defined as follows.

- In the life cycle a sexual phase is followed by three phases of asexual multiplication.
- The sexual phase and first asexual phase occur in *Anopheles* mosquitoes.
- The second and third asexual phases occur in a vertebrate species: the second in the liver or other tissues, the third in the blood. The third phase may be repeated many times.
- Gametocytes are produced in erythrocytes of the vertebrate host, and mature into male and female sex-cells (gametes) in the stomach of a mosquito which has fed on the host.
- A specific malarial pigment is present in some stages of the parasite. This pigment, haemozoin, is produced during the digestion of haemoglobin by the parasite.

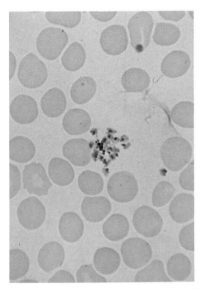

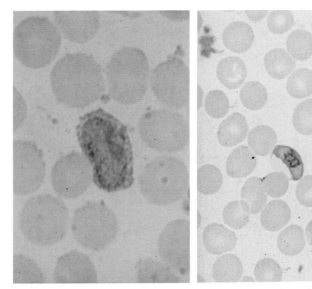

39. Merozoites escaping from a ruptured
schizont.   *World Health Organization*

40. Gametocytes of *Plasmodium vivax* (left) and *Plasmodium falciparum*
(right), as they appear in a stained light-microscope preparation.

World Health Organization

During most of the life cycle the organism is haploid (it has only one
set of chromosomes). It is diploid with two sets of chromosomes for
a brief period in the sexual phase: the zygote and ookinete stages.
Reduction division occurs as the first nuclear division after
fertilization.

Plasmodium is an ancient and successful genus. More than 125
species are known. Most infect birds, but others are found in reptiles
such as snakes, lizards, and turtles. Among mammals *Plasmodium*
species occur in primates and rodents. Other mammalian families
appear to be free of malaria, notably carnivores, ungulates, and
elephants. *Plasmodium* seems to be associated with forest fauna.

Some ape and monkey species of *Plasmodium* occasionally infect
humans naturally, or can be induced to do so experimentally. Such
zoonotic forms of malaria are thought not to be important, but little
is known about them.

Some species of *Plasmodium* are very valuable as laboratory models
of human malaria. Thus Ross studied *Plasmodium relictum* in Indian
sparrows to confirm his theory of the life cycle. Other examples are
Plasmodium gallinaceum in chickens, *Plasmodium berghei* in mice and
rats, and *Plasmodium cynomolgi* in monkeys.

2.5   The malaria mosquito, *Anopheles*

Mosquitoes of the genus *Anopheles* are common in most temperate
and tropical countries, provided only that there are suitable breeding
sites. They are active at night, sheltering in shaded humid places
during the day. They are relatively small insects, up to 8 mm. long.
They fly more quietly and bite more subtly than other blood-sucking
insects. They are easy to overlook.

The distinctive features of an *Anopheles* mosquito are shown in
(Fig. 41). *Anopheles* adults have dark-spotted or dappled wings. Their
posture when resting or biting is distinctive (Fig. 42): head down,
the body at an angle, and the hind legs raised. Long palps flank the
proboscis in both sexes.

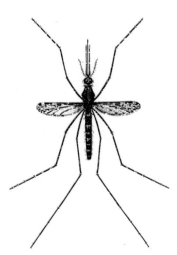

Anopheles gambiae, the most important vector of malaria in Africa.   *Prof. J. D. Gillett*

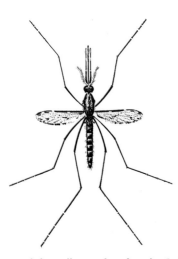

Anopheles wellcomei, found in the Sudan, named after Sir Henry Wellcome.   *Prof. J. D. Gillett*

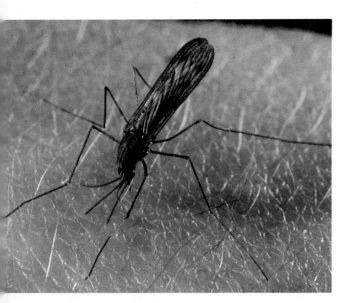

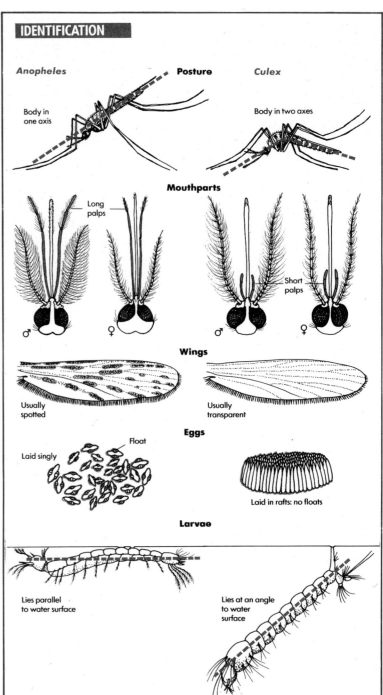

41. The identification of *Anopheles*.   *Trustees of the Wellcome Trust*

42. A female *Anopheles* sucking blood, showing the proboscis, palps, dappled wings, and distinctive posture.   *London Scientific Films Ltd*

Anopheles larvae prefer clean water, fresh for most species but brackish for some. They live just beneath the surface with the body about horizontal. They breathe through a tail pipe and filter water for their food, which is mostly algae and bacteria. If disturbed, they dive to the bottom and hide.

Anopheles pupae do not feed (Fig. 43). They rest just under the surface and breathe through two trumpet-like structures on the head, while the mature insect develops within. They too dive and hide if disturbed. When mature they rise to the surface and split, and the adult mosquito climbs free.

Anopheles prefer a rural habitat but adult mosquitoes can fly several kilometres in one night, invading towns and laying eggs in wells, drains, and ponds. *Anopheles* species differ greatly in behaviour. Some prefer to bite humans, and are called anthropophilic: others feed for preference on animals, and are zoophilic. Mosquitoes which will enter or inhabit houses are called endophilic, and endophagic if they normally feed indoors. Exophilic mosquitoes live outdoors, and are exophagic if they normally feed there.

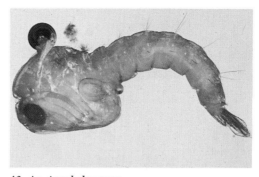

43. An *Anopheles* pupa.

Armed Forces Institute of Pathology, Washington

The life cycle of *Anopheles* (Fig. 45)

From egg to a new adult *Anopheles* takes 7–21 days, depending on temperature. Mating is the first activity of the newly-hatched adult. The female copulates once only, storing sperm for all subsequent egg production. Indeed she cannot mate again, because the final act of the male is to inject a sealing substance which blocks the passage of sperm from any subsequent copulations. Female *Anopheles* need a blood meal for each batch of eggs, 100 or more, produced at two- or three-day intervals. They can feed on nectar or fruit juices, but without blood cannot produce eggs. The factors in blood required for egg production are unknown.

In warm wet conditions the reproductive potential of the female *Anopheles* is enormous, and is a major factor in the success of the malaria parasite.

Development of *Plasmodium* in the mosquito takes 8–10 days at least, so, after taking a blood meal containing parasites, the female has to survive four or five egg-production cycles before she becomes infectious. Mosquitoes transmitting malaria are at least middle-aged. The female may live one month or more in favourable circumstances, but her life-span is often much shorter. The long duration of sporogony in the poikilothermic mosquito is the one point of vulnerability in the life cycle of *Plasmodium*. It explains why malaria transmission increases rapidly as the mean ambient temperature rises.

Male *Anopheles* feed on fruit juices and do not transmit malaria.

An *Anopheles* species which is able to transmit malaria efficiently will usually have the following features:

● It breeds prolifically. The population density is high (Fig. 44).
● It prefers to feed on man: it is anthropophilic.
● It lives and feeds indoors if it can: it is endophilic and endophagic.
● Infected females have a good chance of surviving long enough for the parasite to mature. At least 70 per cent must survive each day if more than 1 per cent are to survive the ten-day development time of *Plasmodium* at an ambient temperature of 30°C.

44. In warm wet conditions mosquito numbers can be enormous. The picture shows one night's catch by a mosquito trap.

Mosquito Research & Control Unit, Cayman Islands

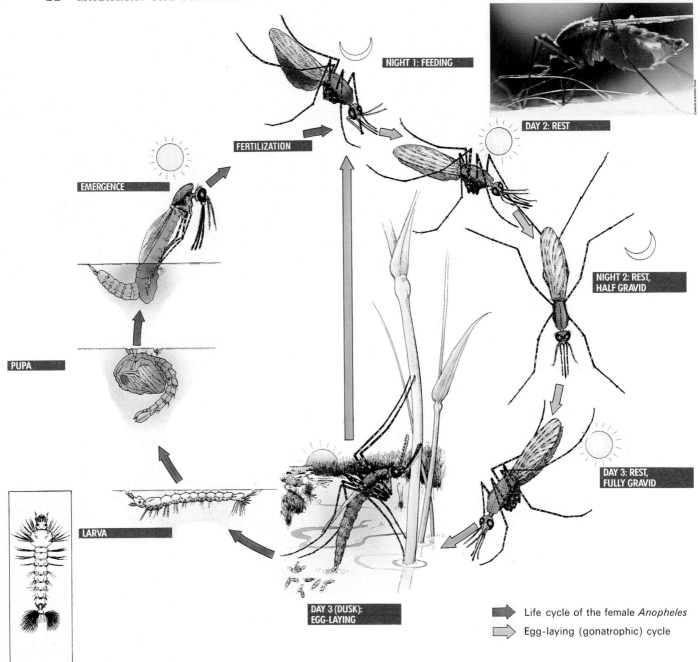

NIGHT 1: FEEDING

DAY 2: REST

FERTILIZATION

EMERGENCE

NIGHT 2: REST,
HALF GRAVID

PUPA

DAY 3: REST,
FULLY GRAVID

LARVA

DAY 3 (DUSK):
EGG-LAYING

➡ Life cycle of the female *Anopheles*

➡ Egg-laying (gonatrophic) cycle

45. The life cycle of *Anopheles*.

Night 1 : feeding.
The female *Anopheles* has to forage and take a blood meal. If she fails she rests during the day and forages again the next night.

Day 2: rest. During the day *Anopheles* mosquitoes rest in cool, shaded, humid places. A fed *Anopheles* begins to digest the blood and her eggs start to develop.

Night 2: rest, half gravid. The insect rests, blood digestion and egg production continue. She may leave the house or change her rest site.

Day 3: rest, gravid. Still in a cool, shaded, humid place. Egg production is completed and the female is fully gravid.

Day 3: dusk. The gravid female flies to a suitable water collection and lays 50 to 150 eggs. She will then forage for another blood meal to start the cycle again.

Larva. The eggs hatch in 2 to 3 days. The larvae or 'wrigglers' feed by filtering algae and other material from the water. In favourable conditions they grow rapidly, passing through 3 moults in as many days.

Pupa. After the third moult the larva feeds and grows again, then becomes a mobile pupa. There is no feeding at this stage. The pupa breathes through 2 air-trumpets

while development of the adult proceeds internally.

Emergence. After 2–3 days the adult emerges. The pupa splits and the soft limp adult climbs out. It needs to dry and harden for some time before it can fly.

Fertilization. Mating is the first activity of the young adult *Anopheles*. The males in many species form mating swarms, usually at twilight. After copulation a mating plug is formed in the female's genital passages, probably from a secretion deposited by the male at the end of copulation.   *Trustees of the Wellcome Trust*

2.6   Malaria afflicts most of the world's people

One hundred million cases of malaria annually (Fig. 46). That is the official estimate from the World Health Organisation, but the true figure is certainly much greater. Notification of cases is unreliable in many tropical areas, and many African countries do not report malaria statistics to the World Health Organisation. The world total of deaths each year from malaria is estimated to be between 1 and 2 millions, the majority being children under five, and almost all attributed to *Plasmodium falciparum*.

Endemic areas are those in which malaria transmission takes place and new cases occur over a number of years. Indigenous malaria transmission requires an average temperature exceeding 15°C for at least one month in the year. It has occurred at 64° North (Archangel in the USSR), and at 31° South (Cordoba in Argentina). Areas free from malaria occur within these limits: it cannot survive above 3000 metres anywhere in the world, and its maximum altitude declines rapidly with distance from the equator. Dry and desert areas are also unfavourable, so malaria is a focal disease over a large part of the earth (Fig. 47).

Plasmodium vivax has the widest geographical range, occurring in temperate and tropical zones. *Plasmodium falciparum* predominates in Africa and much of South-East Asia, but *Plasmodium vivax* is the more common in Southern Asia and Central America. Black Africans tend to resist *Plasmodium vivax* infection, because the Duffy blood-group antigen is rare in this population, and this is the erythrocyte molecule to which *Plasmodium vivax* merozoites bind. *Plasmodium*

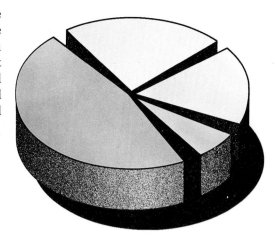

Total World population: 4751 million.

In areas where malaria never existed or disappeared without antimalarial campaigns: 1311 million (28%).

In areas from which malaria has been eradicated in the past 50 years: 776 million (16%).

In malarious areas where control measures continue: 2266 million (48%).

In malarious areas with no control measures: 398 million (8%).

46. Pie diagram illustrating the proportions of the world's population subject to different levels of the threat of malaria. *Trustees of the Wellcome Trust*

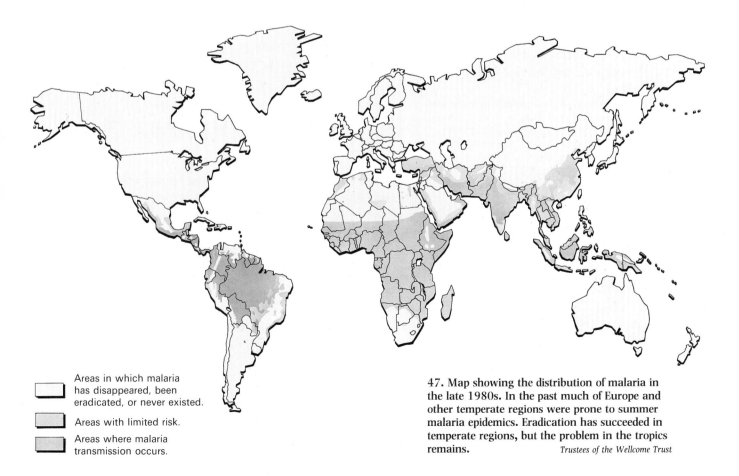

Areas in which malaria has disappeared, been eradicated, or never existed.

Areas with limited risk.

Areas where malaria transmission occurs.

47. Map showing the distribution of malaria in the late 1980s. In the past much of Europe and other temperate regions were prone to summer malaria epidemics. Eradication has succeeded in temperate regions, but the problem in the tropics remains. *Trustees of the Wellcome Trust*

48. Malaria is endemic in all equatorial rain forests: this is in Papua New Guinea.

Transmission is continuous and often intense, with no dry season to interrupt.

Dr. J. D. Charlwood

49. When the rains come, epidemic malaria may come too.

Epidemic malaria is a serious problem in savannahs and other areas bordering endemic zones. Mountain areas within endemic zones may show a similar pattern of transmission. Malaria comes with the rains, in the season of greatest food production. Malaria is a cause of famine.

Food and Agriculture Organization

50. Tropical towns do not escape malaria.

Water supply, sewers, and drainage must be carefully managed to limit mosquito reproduction.

TALC

ovale partly substitutes for *Plasmodium vivax* in Africa, but is an unusual species even there. *Plasmodium malariae* has much the same range as *Plasmodium falciparum*, but is much less frequent.

Malaria has been eradicated from most temperate areas, but cases in travellers are increasing. Throughout the tropics malaria causes anaemia, illness, and death on an enormous scale.

No vaccine giving long-term protection against malaria is available or in prospect. Large-scale use of antimalarial drugs for prevention causes many problems with drug safety, and leads to the emergence of drug resistance. The *Anopheles* mosquito vector is the only practicable target for malaria control. When properly applied mosquito control is usually found to be cost-effective, but still there are many difficulties.

- Vast areas must be treated for an indefinite time.
- The mosquito soon becomes resistant to insecticides.
- People often refuse repeated spraying of their houses, and treatment of sacred buildings may be forbidden.
- Trained personnel are in short supply.
- The environment may be damaged: no insecticide is specific for mosquitoes.
- The total cost is enormous.

Malaria control is like the supply of clean water. It is vital, but tedious and undramatic: 'mud, blood, and sweat'. For indigenous adults in endemic areas acute attacks may be no more troublesome than the common cold, and the need for control may be questioned by non-medical people.

The great hopes for malaria eradication in the 1960s have not been realized. We now realize that no technical assault on the malaria problem will succeed without a concomitant improvement of the social and economic conditions of populations at risk.

2.7   Human communities and malaria

The threat of malaria varies according to the environment and the immunity of individuals at risk.

In tropical areas with rainfall throughout the year malaria transmission is continuous, and indigenous people are exposed constantly to infection (Fig. 48). The disease kills many young children, but survivors develop premunition. This is a state of partial immunity which limits *Plasmodium* multiplication but does not clear the body of parasites. The infected person and the parasite achieve a fragile equilibrium. Repeated infections are necessary to maintain premunition, but symptoms are usually minor. People with premunition transmit the disease for long periods, because parasites circulate in their blood for much of the time. Visitors to such regions are at serious risk. People from endemic areas lose premunition in a few months if exposure to malaria ceases, and are then again at risk when re-exposed.

In drier tropical areas where rainfall is sporadic or seasonal malaria is transmitted only during wet seasons, and the threat may vary greatly from year to year (Fig. 49). Case numbers may be very high during an epidemic, and adults and children die. At other times the disease may be rare. Adults do not transmit the disease without having symptoms, and are not infected constantly to maintain premunition. Visitors are at relatively low risk, except during epidemics.

Tropical towns do not escape malaria (Fig. 50), but the relation to the seasons and rainfall is disturbed. Water-supply systems, waste-water and rain drainage, and sewage disposal must each be carefully managed to prevent mosquitoes from breeding.

Most endemic areas are in the tropical rain forest regions. Epidemic areas are drier, adjacent to the tropics, or on high ground within an endemic region. Until recently epidemic malaria occurred in much of Europe during the summer and autumn, spreading in some years as far north as Archangel. Today there is no malaria transmission anywhere in Europe, thanks to effective eradication campaigns and continuing vigilance.

Although malaria is no longer transmitted in Europe, the number of cases diagnosed is increasing (Fig. 51). This is a consequence of increased travel, especially by air. Such cases of 'imported malaria' are at risk of misdiagnosis and delayed treatment, sometimes with fatal results.

2.8 Chronic malaria causes serious complications

Debilitation, malnutrition, enlargement of the spleen, and anaemia are the expected complications of repeated attacks of malaria (Fig. 52). In some endemic areas the majority of schoolchildren will be found to have enlargement of the spleen. The risks of chronic malaria are greatest for children, whose growth and development may be seriously and irreversibly impaired. These complications are predictable, given the nature of the disease, and will occur in all patients.

In any population, some individuals will be found who react to repeated or persistent *Plasmodium* infection in unusual ways, probably related to an idiosyncrasy in their immune responses. In such people one of three complications may occur.

- Tropical splenomegaly syndrome, in which the spleen becomes massive. This usually occurs in young adults, and regresses during prolonged antimalarial treatment. It sometimes runs in families.
- Nephrotic syndrome, a kidney disease causing fluid retention and oedema (Fig. 53). This is common in African children, and is associated with an altered immune response to *Plasmodium malariae*.
- Burkitt's lymphoma, a very malignant tumour of lymphoid tissues. It typically presents as a large tumour of the jaw (Fig. 54), ulcerating in the mouth and eroding gums and teeth. Invasion of the orbit threatens the eye. Abdominal and central-nervous-system disease is common. The tumour is probably caused by Epstein-Barr virus. This is familiar as the cause of glandular fever, but it can also cause malignancy when the immune system is altered or suppressed (Fig. 55).

Thus, in endemic areas malaria brings to children the immediate threat of lethal illness; the prospect of development impaired by infection, anaemia, and malnutrition; and the distant risk of serious complications, which may develop when their defensive immunity is suppressed by malaria. Where Burkitt's lymphoma occurs in Africa it may be the commonest cancer in children: indeed it can be

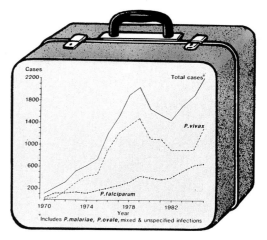

51. Cases of 'imported malaria' in the UK 1970–1985. *Trustees of the Wellcome Trust*

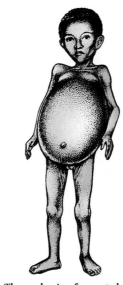

52. The cachexia of repeated malaria in an African child. *Trustees of the Wellcome Trust*

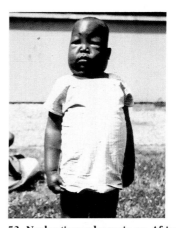

53. Nephrotic syndrome in an African child.
There is oedema of the whole body. This disease appears to be especially common as a complication of quartan malaria (*Plasmodium malariae*).
Prof. P. D. Marsden

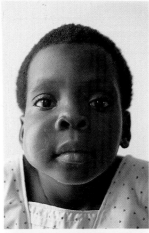

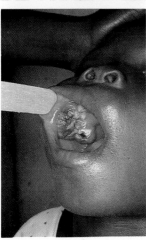

54. Burkitt's lymphoma in a nine year old Ghanaian girl.

The tumour is in the maxilla, and eroding the gum and cheek lining. There was a large ovarian tumour as well. Treatment with modern cytotoxic drugs has a good chance of success, but is too expensive for many tropical countries.

Trustees of the Wellcome Trust

commoner than all other childhood cancers added together. It responds well to modern cytotoxic drug therapy, but this itself has hazards and possible long-term complications. In any case the very high cost of treating even one patient precludes its use in most tropical areas.

Genetic blood diseases and malaria

Chronic malaria is more than a threat to individuals. Serious genetic blood disorders increase in populations stressed by endemic malaria. Sickle-cell disease and thalassaemia are the most common. Each disease is caused by the presence of abnormal haemoglobin in the patient's erythrocytes. In sickle-cell disease it is haemoglobin S (HbS). In thalassaemia one of the two types of sub-unit in normal haemoglobin is not produced.

The inheritance of the sickle-cell gene is shown in Fig. 56. A normal individual produces haemoglobin (AA) on the instructions of two sets of genes, one set from each parent. If one parent has transmitted the abnormal gene for haemoglobin S then the individual will have a mixture of normal haemoglobin and haemoglobin S. This is the sickle-cell trait (AS), and the individual will appear normal unless specially tested. If genes for haemoglobin S are inherited from both parents then the child has sickle-cell anaemia (SS). This is a devastating illness, and the child may die young. Survivors are permanently severely anaemic; have bones damaged or distorted by the overgrowth of red bone marrow, a natural response to prolonged severe anaemia (Fig. 57); and are prone to painful and destructive thrombotic crises.

The sickle-cell trait (AS) protects against *Plasmodium falciparum* malaria. In Africa and other endemic regions children with sickle-cell trait (AS) are more likely than normal children (AA) to survive to reproductive age. The increased survival of children with sickle-cell trait (AS) exceeds the deaths of children with sickle-cell anaemia. The advantage of (AS) is greater than the disadvantage of (SS), so the number of individuals carrying the disease (AS) increases. Any gene which increases the net reproductive potential of its carriers will be strongly selected for, so long as the circumstances which cause its advantage persist. If the stress of *Plasmodium falciparum* on a population is reduced or removed then the sickle-cell gene is detrimental to all its carriers, and will tend to die away.

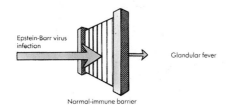

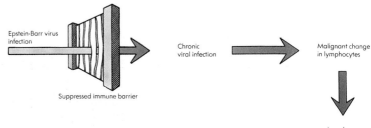

55. When malaria depresses immunity the Epstein-Barr virus may cause Burkitt's lymphoma.

Immune suppression by malaria allows the Epstein-Barr virus to produce malignancy in lymphocytes, and Burkitt's lymphoma may develop.

Trustees of the Wellcome Trust

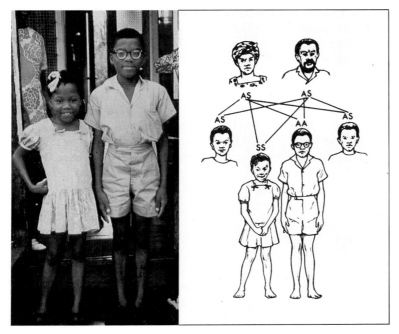

56. Sickle-cell genes and malaria.

Twins aged eight. The girl has sickle-cell anaemia, HbSS. The boy has normal HbAA.

Prof. G. R. Sergeant

AA. Normal individuals. The erythrocytes contain only normal haemoglobin, HbA.
AS. The sickle-cell trait. The erythrocytes contain a mixture of HbA and HbS. Individuals are moderately anaemic but usually healthy.
SS. Sickle-cell disease.   *Trustees of the Wellcome Trust*

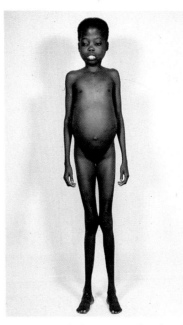

This genetic effect is called balanced polymorphism (Fig. 58). Similar genetic mechanisms probably operate in other endemic areas to cause the genes for thalassaemia and other blood diseases to accumulate in populations exposed over many generations.

2.9   Problems of cure

Antimalarial drugs

Drugs commonly used to treat malaria are summarized in Fig. 59. Quinine in the form of *Cinchona* bark came to Europe in the seventeenth century. Its specific value in intermittent fevers was recognized in 1666 by Morton and Sydenham, and confirmed by

57. The body-habit associated with sickle-cell anaemia, but fully developed only in a minority.

The limbs are elongated, the skull bossed, and the teeth and jaws protrude.   *Prof. P. D. Marsden*

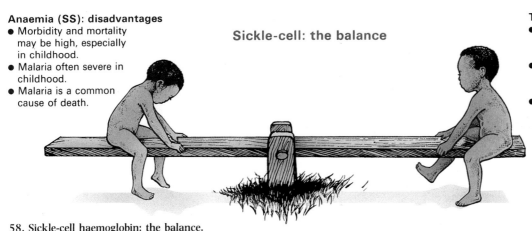

Anaemia (SS): disadvantages
● Morbidity and mortality may be high, especially in childhood.
● Malaria often severe in childhood.
● Malaria is a common cause of death.

Sickle-cell: the balance

Trait (AS): advantages
● Essentially healthy individuals with normal life span.
● Fewer and less severe attacks of malaria in childhood.
● Rarely die from malaria.

58. Sickle-cell haemoglobin: the balance.

The disadvantages of sickle-cell anaemia: morbidity and mortality are high, especially in children; malaria is often severe or fatal.
The advantages of sickle-cell trait: individuals are generally healthy although anaemic; the risk of death in childhood from *falciparum* malaria is much reduced.

Trustees of the Wellcome Trust

Torti in 1712. Chemotherapy for malaria was available two centuries before the parasite was discovered.

All other antimalarials have come from pharmacological research in this century, often stimulated by the medical problems of war. The research effort has been immense and frustrating. During the Second World War only the penicillin project had greater investment than the development of antimalarials. The number of chemical compounds synthesized and tested as antimalarials exceeded a quarter of a million, and at least the same number has been tested since 1945. Only a handful have proved suitable for general use.

We know that proguanil, pyrimethamine, and related drugs act by inhibiting folic acid metabolism much more strongly in the parasite than in the host. The development of the parasite is inhibited by these drugs, which arrest maturation, producing large, non-viable organisms (Fig. 60). It is exactly the same effect as folic acid or vitamin B_{12} deficiency has on human erythrocyte maturation.

The mechanism of action of other groups of antimalarials is obscure. Chloroquine forms complexes with the haematin of malaria pigment, causing it to aggregate into coarse granules. Quinine and related quinoline drugs do cause some clumping of pigment, but also produce nuclear and cytoplasmic degeneration (Fig. 61). Some of these drugs can interact with *Plasmodium* nucleic acids by intercalation: the molecules of the drug slide between the stacked bases of the nucleic acid, so distorting its structure.

Recently a new type of antimalarial drug has emerged from studies of an old Chinese remedy for fevers. This is Qinghaosu, derived from the plants of the genus *Artemisia*, which contain chemical substances related to Artemisinine. These compounds are organic peroxides, and may damage the parasite oxidatively.

Most antimalarial drugs are relatively specific for *Plasmodium*, and of little value in treating other infections (the antifolate drugs are

Classification of antimalarial drugs

Group	Compound types	Stage of life cycle affected	Action on cells	Action on malarial pigment
1	Chloroquine Mepacrine 4-Aminoquinolines	All asexual stages	–	Rapid coarse clumping
2	Quinine Mefloquine Primaquine 8-Aminoquinolines	All stages except mature gametocytes of *P. falciparum* (but Primaquine active)	● Degeneration of nuclei ● Vacuolation of cytoplasm	Slow, fine clumping
3	Antifolates: ● Proguanil ● Pyrimethamine ● Sulphonamides	Schizogony	Maturation arrest producing large non-viable parasites	–
4	Sesquiterpenes: ● Artemisinine	Schizogony	–	–

59. A table to show the main groups of antimalarial drugs and some of their characteristics.

an exception). This suggests that *Plasmodium* has special metabolic processes unusual in other organisms.

Three forms of *Plasmodium* occur during human infection, the hepatic phase including hypnozoites, the asexual erythrocytic phase, and gametocytes. Only primaquine (a rather toxic drug) kills all human forms of *Plasmodium*. For this reason a number of special terms were introduced to identify the objectives of therapy in an individual case.

- *Causal prophylaxis* is therapy given to suppress hepatic schizogony.
- *Clinical prophylaxis* is the suppression of erythrocytic schizogony. This is sometimes called suppressive prophylaxis.
- *Suppressive cure* is the eradication of *Plasmodium* by continued clinical prophylaxis.
- *Radical cure* is the eradication of hepatic schizonts, hypnozoites, and erythrocytic schizonts by combined causal prophylaxis and suppressive cure.

Antimalarial drug resistance

Resistance to antimalarial drugs is spreading rapidly in most tropical areas. In South-East Asia, for example, strains of *Plasmodium falciparum* may be resistant to most or all conventional drugs. Drug resistance results from two factors.

- The remarkable adaptability of *Plasmodium*.
- The use of antimalarials for prophylaxis, and for inadequate routine treatment of undiagnosed fevers in endemic areas.

Resistance patterns are shown in Fig. 62, classified in the terms defined by the World Health Organization. The two discriminating factors are initial clearance of parasites from the blood, and later recrudescence.

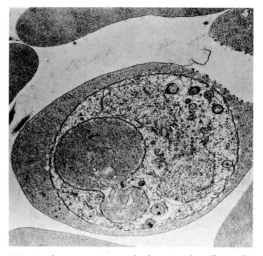

60. An electron micrograph showing the effects of an antifolate drug on an erythrocytic trophozoite. Development has ceased at the stage of early schizogony.   *Prof. W. Peters*

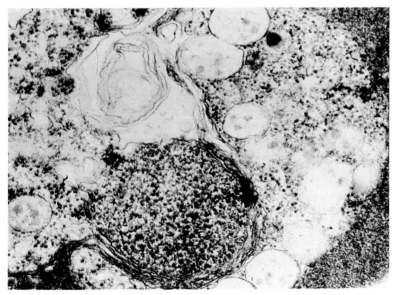

61. An electron micrograph of an erythrocytic trophozoite following treatment with mefloquine (a drug related to quinine). Digestive vacuoles are enlarged and the cytoplasm shows degenerative changes.

Prof. W. Peters/Dr. D. C. Warhurst

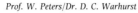

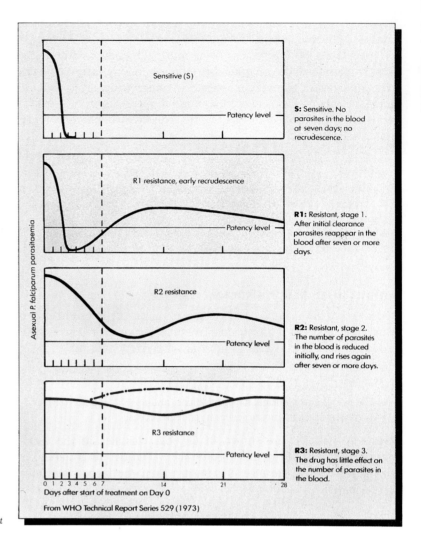

**62. The classification of drug resistance in
Plasmodium.** *Trustees of the Wellcome Trust*

S The strain of *Plasmodium* is sensitive to the drug used. Parasites
   are cleared from the blood within seven days and do not
   reappear.

R1 Resistant, stage 1. After initial clearance parasites reappear
   after seven or more days despite continuing treatment.

R2 Resistant, stage 2. The number of parasites is reduced at first,
   but rises again after seven or more days.

R3 Resistant, stage 3. The drug has little effect on parasite
   numbers.

The remorseless spread of drug resistance is perhaps the most
alarming problem for malaria specialists. From a huge investment
came no more than twenty antimalarial drugs of proven value and
safety. Inappropriate or careless use has squandered much of the
advantage gained. New drugs are appearing, such as Mefloquine and
Qinghaosu, but resistance to them has appeared already.

Quinine remains effective in most areas, but approximately half of
the world's production is used to give a bitter flavour to tonic waters.
This custom exposes the malaria parasite to low concentrations of
this valuable drug: the perfect conditions for the development of
resistance, although none attributable to tonic waters has yet been
proven.

The appearance of drug resistance means increased risk of side-effects or greater expense, or both, as second-line or new drugs have to be used. The disadvantages of second-line drugs and reluctance to lose effective new drugs mean that safe, reliable malaria prophylaxis may no longer always be possible. Personal and domestic precautions against mosquito bites are again increasing in importance in malaria control.

2.10 Problems of control

In principle, malaria can be eliminated from the world. The parasite and the mosquito are vulnerable to public health measures. After 1945 hopes for global eradication were raised by the following observations.

1. Host specificity
For practical purposes man is the only host of human species of *Plasmodium*: there is no reservoir in animals, although some primates can be infected with human species.

63. Collecting mosquitoes resting beneath a thatched roof. *Prof. L. J. Bruce-Chwatt*

2. Vector specificity
Only female *Anopheles* mosquitoes transmit malaria. In some species of *Anopheles* the females are efficient vectors, but in others they are not.

3. Vector behaviour
After each feed the female *Anopheles* rests on a wall or other surface indoors until the blood is digested (Fig. 63). They do not disperse, and control measures can be restricted accordingly.

4. Duration of sporogony
Plasmodium requires 10–30 days to complete its development in the mosquito, depending on the temperature of the environment. During this period the mosquito must take at least four more blood meals with rests between. There is a good opportunity to destroy the vector mosquito before it can transmit the disease.

5. New insecticides
The advent of DDT provided an insecticide that is non-toxic to humans, cheap, and persistent, therefore not needing frequent applications. It seemed an ideal weapon against mosquitoes resting on walls and ceilings, or in roofs (Fig. 64).

64. A spraying team in action.

Meddia, Royal Tropical Institute, Netherlands

6. Further measures
The benefits of the mosquito control could be reinforced by prompt diagnosis and treatment of human cases; population surveys to find and treat carriers (called surveillance); and other measures to reduce the number of breeding sites for mosquitoes.

The global malaria eradication programme

Malaria eradication was feasible, but needed meticulous application worldwide.The costs would be high, but the benefits incalculable. Malaria eradication campaigns in Italy (Fig. 65), Greece, Cyprus, Guyana, and Venezuela succeeded. The strategy and tactics appeared sound.

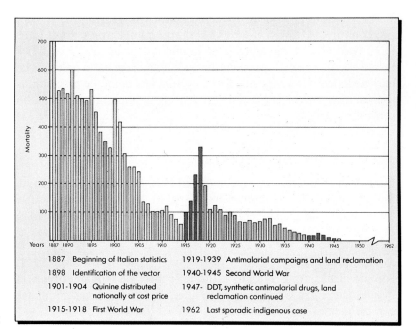

65. The decline of malaria in Italy (from Bruce-Chwatt and de Zulueta (1980)).

Trustees of the Wellcome Trust

1887	Beginning of Italian statistics	1919-1939	Antimalarial campaigns and land reclamation
1898	Identification of the vector	1940-1945	Second World War
1901-1904	Quinine distributed nationally at cost price	1947-	DDT, synthetic antimalarial drugs, land reclamation continued
1915-1918	First World War	1962	Last sporadic indigenous case

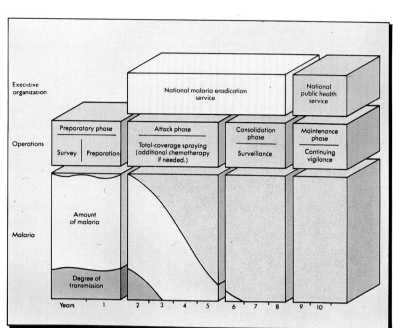

66. The structure of an ideal malaria eradication programme (from Bruce-Chwatt (1985)).

Trustees of the Wellcome Trust

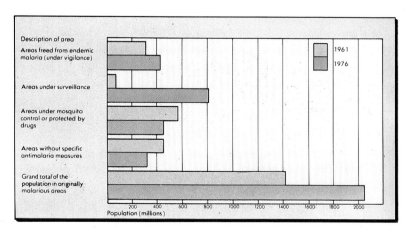

67. The progress of malaria eradication and control between 1961 and 1976, according to the WHO.

Trustees of the Wellcome Trust

In 1955 the Eighth World Health Assembly announced a programme for global malaria eradication. All countries affected by malaria would mobilize against the disease, encouraged and supported by the World Health Organization, modifying a basic plan for each national eradication campaign (Fig. 66).

Mosquito resistance to DDT and soon to other insecticides emerged during the early campaigns, but had taken at least five years to develop. It was expected that malaria would not reappear if transmission was interrupted and all cases treated for three years. Therefore eradication programmes would need energetic and thorough operation, but for a limited period.

Against this background the World Malaria Eradication Campaign began. It was a notable enterprise, well researched, well planned, and well funded.

By 1970 malaria eradication had succeeded in the whole of Europe, the USSR, most of North America, several Middle Eastern countries, large parts of South America and the Caribbean, Australia, Japan, Taiwan, and Singapore. It was a remarkable public health achievement (Fig. 67). But it was not enough.

Seasonal or endemic transmission continued in many tropical countries. Difficulties in implementing eradication programmes were compounded by further problems developing during the 1970s: the energy crisis, inflation, and the rapid increase in mosquito resistance to insecticides and parasite resistance to drugs. Political instability and war impeded national eradication projects. Malaria control programmes faltered, while human numbers at risk increased explosively. Between 1973 and 1977 the number of cases of malaria reported to the WHO increased 2.5 times. The rapid expansion of air travel added to the numbers at risk, and cases of imported malaria increased.

At present world malaria case numbers change little: improvements in some areas balance deterioration in others. Malaria today affects more people than it did in 1960, and there is no prospect of a substantial improvement. Vaccines are unlikely to provide a quick technical fix. Improved standards of living and reduction in poverty will be essential if eradication is to succeed in many countries.

3 *Plasmodium*: the parasite

The past ten years have seen a rapid increase in our knowledge of the detailed structure of *Plasmodium*. The culture of *Plasmodium falciparum* was finally achieved in 1976, after many years of trial and failure. This success made possible dramatic growth in the cell-biology of *Plasmodium*, by making parasites freely available for laboratory study, especially by electron microscopy. Other human species of *Plasmodium* have proved even more difficult to culture, but the techniques are slowly improving. Mass culture of infected *Anopheles* is possible in a few centres, notably in Nijmegen. This provides material for the study of sporogony, and sporozoites to initiate the liver stages, but the cost of the total security necessary to contain infected insects is a serious obstacle. The rodent parasite, *Plasmodium berghei*, has made possible detailed study of the liver stages in rats.

It is now possible for the first time to display a complete sequence of electron micrographs showing the ultrastructure of each stage in the life cycle. The biochemical pathways of *Plasmodium* are still poorly understood, but it is possible to analyse the metabolism of the erythrocytic stages. The new techniques of molecular genetics are already yielding exciting results, and facilitating the search for possible *Plasmodium* vaccines. This new knowledge is the basis of this chapter.

Review of the life cycle of *Plasmodium*

The life cycle of *Plasmodium* is summarized in Figs. 68 and 69. It has four phases, one sexual and three asexual. Each phase ends with

68. The life cycle of *Plasmodium*.
Tabular summary of life cycle.

Life cycle: summary

Phase of life cycle:	Occurs in:	Name of phase	Growth form beginning phase	Invasive form completing phase	Multiplication during phase
1 Sexual	Mosquito: stomach cavity	Gamogony & syngamy	—	Ookinete	1
2 Asexual (1)	Mosquito: stomach wall	Sporogony	Sporocyst	Sporozoite	10 000 x
3 Asexual (2)	Human: liver cell	Hepatic schizogony	Hepatic trophozoite	Merozoite	10-30,000 x
4 Asexual (3)	Human: red blood cell	Erythrocytic schizogony	Erythrocytic trophozoite	Merozoite	8-32 x (repeated)

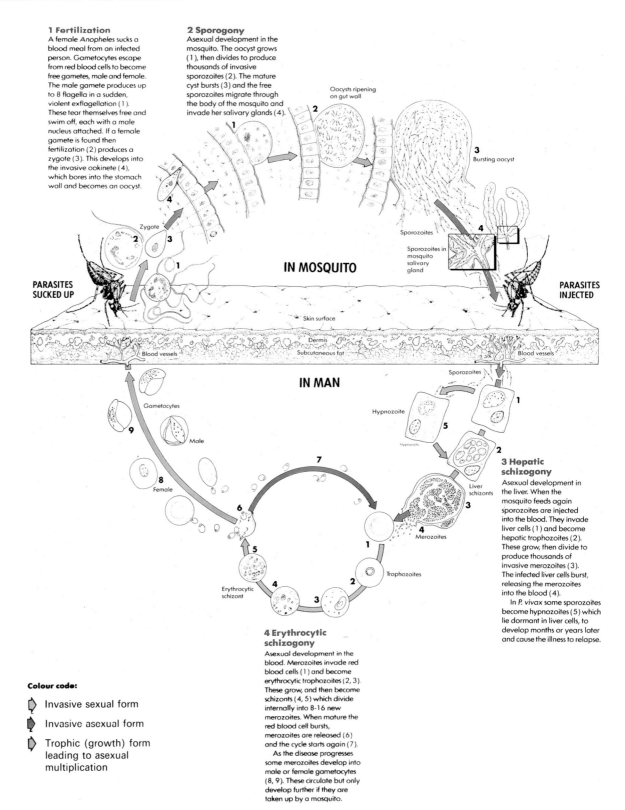

1 Fertilization
A female *Anopheles* sucks a blood meal from an infected person. Gametocytes escape from red blood cells to become free gametes, male and female. The male gamete produces up to 8 flagella in a sudden, violent exflagellation (1). These tear themselves free and swim off, each with a male nucleus attached. If a female gamete is found then fertilization (2) produces a zygote (3). This develops into the invasive ookinete (4), which bores into the stomach wall and becomes an oocyst.

2 Sporogony
Asexual development in the mosquito. The oocyst grows (1), then divides to produce thousands of invasive sporozoites (2). The mature cyst bursts (3) and the free sporozoites migrate through the body of the mosquito and invade her salivary glands (4).

Oocysts ripening on gut wall

Bursting oocyst

Sporozoites

Sporozoites in mosquito salivary gland

IN MOSQUITO

Zygote

PARASITES SUCKED UP

PARASITES INJECTED

Skin surface

Dermis

Subcutaneous fat

Blood vessels

Blood vessels

IN MAN

Sporozoites

Gametocytes

Male

Hypnozoite

Female

Hypnozoite

Liver schizonts

3 Hepatic schizogony
Asexual development in the liver. When the mosquito feeds again sporozoites are injected into the blood. They invade liver cells (1) and become hepatic trophozoites (2). These grow, then divide to produce thousands of invasive merozoites (3). The infected liver cells burst, releasing the merozoites into the blood (4).
     In *P. vivax* some sporozoites become hypnozoites (5) which lie dormant in liver cells, to develop months or years later and cause the illness to relapse.

Merozoites

Trophozoites

Erythrocytic schizont

4 Erythrocytic schizogony
Asexual development in the blood. Merozoites invade red blood cells (1) and become erythrocytic trophozoites (2, 3). These grow, and then become schizonts (4, 5) which divide internally into 8-16 new merozoites. When mature the red blood cell bursts, merozoites are released (6) and the cycle starts again (7).
     As the disease progresses some merozoites develop into male or female gametocytes (8, 9). These circulate but only develop further if they are taken up by a mosquito.

Colour code:

Invasive sexual form

Invasive asexual form

Trophic (growth) form leading to asexual multiplication

69. The life cycle of *Plasmodium* showing the sequence of four phases.

The sexual and first asexual phase take place in an *Anopheles* mosquito, the second and third asexual phases occur in a vertebrate host, man in this case. *Plasmodium* alternates between hosts, and in its asexual phases alternates between intracellular growth forms and extracellular invasive forms. The third asexual phase is repeated many times. Phases in the mosquito are above the horizontal line, in man below.

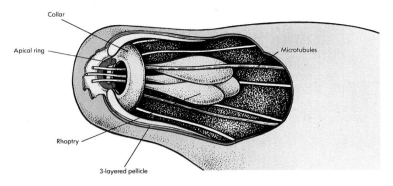

Collar

Apical ring

Microtubules

Rhoptry

3-layered pellicle

70. Diagram of the apical complex.
Microtubules forming the cytoskeleton of an ookinete, and the essential features of the apical complex, the invasive apparatus distinctive of *Plasmodium* and similar parasitic protozoa.

Trustees of the Wellcome Trust

the production of an invasive form, which is extracellular, mobile, and able to enter host cells before the succeeding phase is established. The three asexual stages each begin with an intracellular growth form. This enlarges rapidly before maturing into invasive forms by a process of internal segmentation.

Invasive forms are produced at the end of each phase. These vary in size and shape between phases, but have common structural features, including the following:

- an apical complex (Fig. 70), the apparatus for invasion of host cells, including an apical ring, small vesicles (micronemes), and duct-like structures (rhoptries);
- a thick three-layered pellicle; and
- a cytoskeleton of microtubules, maintaining the pear shape of the organism and contributing to its motility.

In asexual phases the parasites feed and grow rapidly, then divide to produce invasive forms. The division process is similar in each asexual phase.

- The cell nucleus divides repeatedly.
- The progeny nuclei each become associated with the cell membrane, or with deep invaginations from it dividing the parasite cytoplasm into a labyrinth of clefts.
- An apical complex develops in the membrane associated with each nucleus, and then the pellicle and cytoskeleton mature to complete each new invasive individual.
- The residual structure breaks down, releasing the progeny to invade new cells and start a new cycle.

3.1 *Plasmodium* ultrastructure

The cells of *Plasmodium*, like all cells, contain various sub-units collectively called organelles. Each type of organelle has one or more specialized metabolic functions. The size, number, and prominence of different organelles vary during the life cycle, according to changing functional requirements at each stage. In Fig. 71 the more important organelles are identified in a cross-section of a malaria parasite growing in a liver cell. The liver cell is much more ordered and structured than the parasite, at least at this level of magnification, but a number of common organelles can be seen in each. Reaction to the parasite is minimal, and there appears to be little distortion of the hepatocyte cytoplasm, at least at this stage of the parasite's development. It seems to sit in the hepatocyte like a native structure.

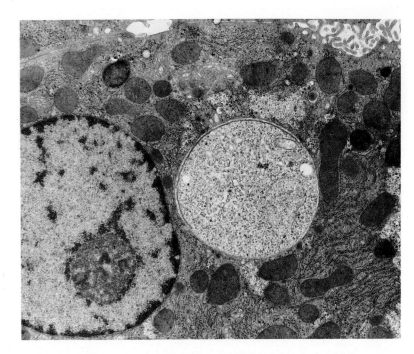

71. A trophozoite in a hepatocyte.

An electron micrograph of a section through a rat liver cell containing a young trophozoite of *Plasmodium berghei* : the parasite causing rodent malaria. When mature the parasite will be several times its present diameter. Peripheral vacuoles and cytoplasmic clefts show that schizogony has begun.

BC: bile caniculus; GV: Golgi vesicles; BB: brush border or sinusoid; PV: parasitophorus vacuole; PM: parasite mitochondrion; PN: parasite nucleus; ER: endoplasmic reticulum with ribosomes; C: cleft in parasite cytoplasm; V: vacuole; M: mitochondrion; N^1: nucleus; N^2: nucleolus; NP: nuclear pore.

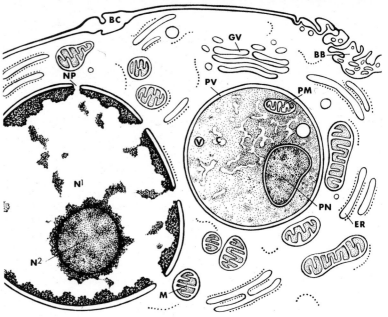

The nucleus

The hepatocyte nucleus is clearly bounded by a double membrane interrupted by several nuclear pores. Much of the nucleus is occupied by nucleoprotein, a complex of deoxyribonucleic acid (DNA) and protein. DNA is the genetic material containing information to control and programme the cell. Within the nucleus is a nucleolus, containing mostly ribonucleic acid (RNA). RNA is transcribed from DNA, and is the vehicle taking coded instructions from the DNA to the cytoplasm. The liver-cell nucleus is diploid—it contains two complete sets of genetic information. At cell division the nucleoprotein is seen to be arranged in 46 chromosomes: 22 pairs plus 2 sex chromosomes.

The parasite nucleus is not so well defined, but it too has a double membrane interrupted by pores. There is no nucleolus. It is haploid, containing one set of genetic information. Chromosomes are not formed at cell division, but there are a number of distinct DNA molecules carrying the genetic information. When schizogony begins the nuclear material is replicated thousands of times, and each new genetic unit moves into a developing merozoite.

The cytoplasm

Mitochondria abound in the cytoplasm of the hepatocyte, but only one is visible within the parasite. Mitochondria may be subdivided by cristae, an indication that they are generating energy oxidatively. Mitochondria are self-replicating, and each stage in the life cycle of *Plasmodium* must carry at least one. Stages which generate their energy non-oxidatively must still contain one rudimentary mitochondrion per cell, to transmit to subsequent stages.

Stacked membranes of the endoplasmic reticulum fill much of the remaining cytoplasmic space in the liver cell. Most are rough endoplasmic reticulum, studded with ribosomes, the apparatus of protein synthesis. The cavities of the endoplasmic reticulum, the smooth endoplasmic reticulum, and Golgi vesicles are concerned with intracellular movement and secretion of protein.

The ribosomes of the parasite mostly lie free. There are several membrane-lined peripheral vesicles, and numerous membrane-bounded clefts are developing in the cytoplasm as schizogony begins. Nucleoprotein and membrane synthesis must dominate the metabolism of the vegetative stages of *Plasmodium*, and the mechanisms of membrane and nucleoprotein reorganization during schizogony are challenging problems for future cell-biologists.

Cell division

Cell division in *Plasmodium* differs in two respects from the usual type of division (mitosis) seen in cells with nuclei, in which multiplication of the genetic material (nucleoprotein), division of the nucleus, and division of the cytoplasm occur in series.

In *Plasmodium* the nucleoprotein and cytoplasm increase, and then numerous progeny are created in a single parallel segmentation, sharing genetic material and cytoplasm among the new individuals. Chromosomes are not formed, which hinders analysis of the process by making the details of nuclear division difficult to follow in microscopical sequences.

Microtubules and flagella

Invasive forms and gametocytes of *Plasmodium* have a cytoskeleton of microtubules. These give a distinct shape to these forms, and contribute to movement. Microtubules are assembled from unit molecules of a protein called tubulin (Fig. 72). The assembly is dynamic, requiring energy, and microtubules can form and disappear rapidly. From disassembled microtubules free tubulin molecules are available for re-assembly in a new microtubule. Guanosine triphosphate (GTP) is the usual energy source for microtubule assembly, and the process is enhanced by two classes of accessory protein; the MAPs (microtubule-associated proteins), and the tau proteins.

Microtubular structure

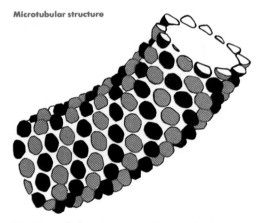

72. Microtubular structure: a diagram to show the assembly of a microtubule from sub-unit tubulin proteins.

Trustees of the Wellcome Trust

NM: 2-layered nuclear membrane.
SF: spindle fibres.
IN MTOC: intranuclear part of MTOC.
MTOC: microtubular organizing centre.
M: mitochondrion.

Microtubule assembly originates in specialized areas. The microtubular organizing centre (MTOC) occurs in the cytoplasm. At cell division the MTOC is associated through a nuclear pore with a similar organizing centre within the nucleus, from which the microtubular spindle fibres radiate (Fig. 73).

Flagella are organized from a basal body, or kinetosome. They contain a central bundle of specialized microtubules, arranged in a ring of nine doublets, with two more at the centre (Fig. 74). The beat of the flagellum is generated by cross-linking the microtubules in a variable, transient manner along the length of the flagellum. The explosive polymerization of flagella in the male *Plasmodium* gamete is remarkable, and their disruption of the cell to swim free with an attached nucleus must be unique in cell biology.

Flagellar structure

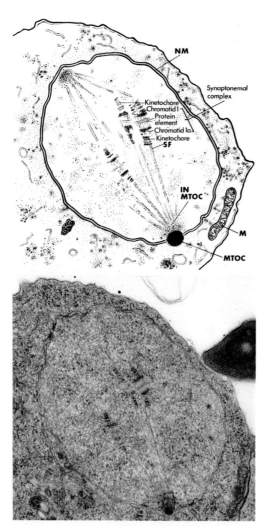

73. An electron micrograph of a section through a maturing zygote of *Plasmodium falciparum*.

It shows the components of the nuclear spindle of the first division of the zygote. This is the reduction division: the two nuclei produced will have the haploid genome usual for *Plasmodium*.

Trustees of the Wellcome Trust
Prof. R. E. Sinden

74. Flagellar structure: a diagram showing the '9 + 2' arrangement of microtubules in a flagellum.

Each of the '9' is a doublet. The microtubules are linked by other proteins which cause the beat of the flagellum. A membrane encloses the central assembly.

Trustees of the Wellcome Trust

3.2 Phase 1. From gametocyte to ookinete Sexual fusion, no multiplication

Activation of gametocytes (Fig. 75)

Dramatic changes occur in gametocytes a few minutes after ingestion by a mosquito. The stimuli to development are uncertain, but gametocytes produced in culture are activated by cooling to below 30°C, or an increase in the alkalinity of the medium. The gametocytes swell rapidly, and discharge their osmiophilic bodies (OB) into the host erythrocyte. Both processes disrupt the erythrocyte membrane, releasing the gametes. The male gamete now produces eight flagella (spermatozoa). During development the DNA of the male gametocyte is replicated three times, so the nucleus of the activated gamete already has eight complete sets of DNA (genomes). Identifiable chromosomes are not formed.

Exflagellation (Fig. 76)

Eight kinetosomes are formed in a cytoplasmic microtubular organizing centre (MTOC) (Fig. 77). Each kinetosome is the base and growing point for the core of a flagellum, or axoneme, which begins to elongate rapidly within the cytoplasm. Each flagellum is made of microtubules, arranged in a '9 + 2' pattern in cross-section.

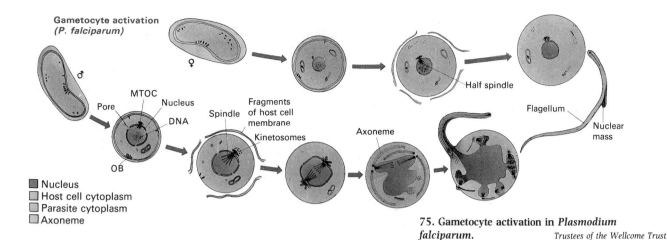

75. **Gametocyte activation in *Plasmodium falciparum*.**

Trustees of the Wellcome Trust

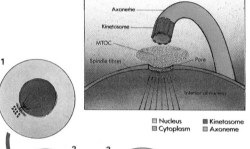
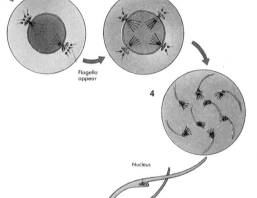
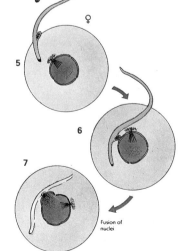

76. **The formation of eight flagella from the male gametocyte, and fertilization.**

1. Spindle fibres extend into the nucleus and attach to each of the eight genomes. Eight kinetosomes are already formed, and polymerization of flagellar axonemes begins.
2. The spindle divides.
3. Now sets of 2 genomes and 2 kinetosomes are arranged around the nucleus in an approximately tetrahedral form. The flagella are extending rapidly.
4. The final division of the spindle associates one genome with each flagellum. The spindle fibres contract and the flagellum breaks free.
5. The head of the flagellum and the membrane of the female gamete fuse.
6. The flagellum drives its front part and the attached male nucleus into the female cytoplasm.
7. The male and female nuclei fuse and the flagellar axoneme depolymerizes.

Trustees of the Wellcome Trust

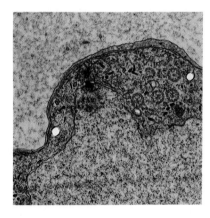

77. **Early exflagellation.**

The microtubular organizing centre (MTOC) is associated with the centriole through a pore in the nuclear membrane. Spindle fibres extend into the nucleus. Numerous microtubules are forming around the MTOC, and within the cytoplasm at least 4 cross-sections of developing flagella are visible. The 9 + 2 arrangement is evident.

Prof. R. E. Sinden

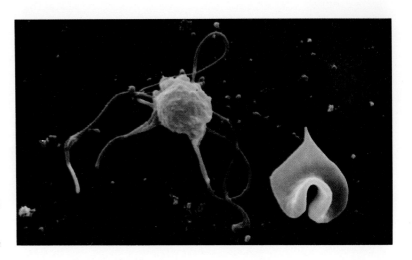

78. A scanning electron micrograph of a male gametocyte at the moment of exflagellation.

The slight swelling at the head of the flagellum, and the bulge along its length where the male nucleus is attached, are visible in one flagellum.

Prof. R. E. Sinden

Fibres of a monopolar mitotic spindle extend into the nucleus from the cytoplasmic MTOC and attach to each of the eight genomes already formed during maturation of the gametocyte. The MTOC divides, and the two halves separate, segregating four genomes and four kinetosomes with developing flagella to each spindle pole. This process is then repeated twice more, producing eight genomes, each attached by spindle fibres to an MTOC containing a kinetosome from which a flagellum is developing.

The flagella begin to beat. The cytoplasm of the gamete appears to boil, then explode, as the flagella tear themselves free, each one dragging a genome with it (Fig. 78). The free flagellum swims off to find a female gamete. It moves in distinctive bursts of beating, and has about one hour of active life.

Fertilization

A flagellum finding a female gamete becomes attached by its head to the female's cell membrane (Fig. 79). The flagellum is inactive while fusion with the cell membrane is completed. Then in a final burst of activity the flagellum drives itself and its male genome into the female cytoplasm. The male and female nuclei fuse, and the remains of the flagellum disintegrate. The product is the fertilized cell, or zygote. This is the only cell in the life cycle of *Plasmodium* which is diploid with two copies of the genome. At the first division of the zygote the DNA content is halved, and all other stages of the life cycle are haploid.

The zygote now differentiates over the next five to ten hours into a cigar-shaped, motile, invasive ookinete (Fig. 80).

Activation of the zygote begins the transformation. Endoplasmic reticulum and Golgi vesicles increase rapidly. Surface membrane components change, and the antigenicity of the organism alters. Next the diploid genome divides with the formation of an intranuclear spindle, the first event in a two-step meiosis. Completion of the process is delayed until sporogony begins in the next phase.

A microtubular organizing centre (MTOC) now appears, associated with one pole of the intranuclear spindle. The MTOC is in turn associated with the components of the apical complex and cytoskeleton, which also begin to develop at this stage. The MTOC controls this process in some way.

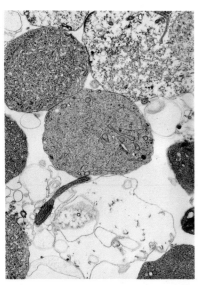

79. The first stage of fertilization. The flagellum is attached by its apex to the membrane of the female gametocyte.

At the point where the head of the flagellum is attached to the female gametocyte the membrane is becoming thickened. *Prof. R. E. Sinden*

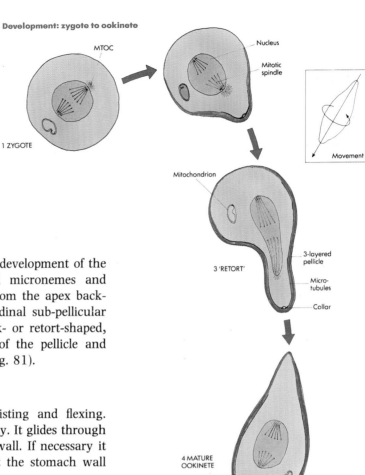

Development: zygote to ookinete

80. A diagram showing the development of an ookinete from a zygote.

The apical complex appears first, and the shape of the ookinete emerges progressively, as the pellicle and microtubular cytoskeleton are formed from the front backwards. *Trustees of the Wellcome Trust*

The anterior of the ookinete is defined by the development of the apical complex: apical rings, the collar, and micronemes and rhoptries. The shape of the ookinete develops from the apex backwards, as the three-layered pellicle and longitudinal sub-pellicular microtubules appear. The early ookinete is flask- or retort-shaped, becoming pear- or cigar-shaped as formation of the pellicle and cytoskeleton is completed at the posterior end (Fig. 81).

The ookinete

The mature ookinete is able to move by twisting and flexing. Microtubules of the cytoskeleton give it this ability. It glides through the blood meal to the cells lining the stomach wall. If necessary it can disrupt red-cells obstructing its passage. At the stomach wall

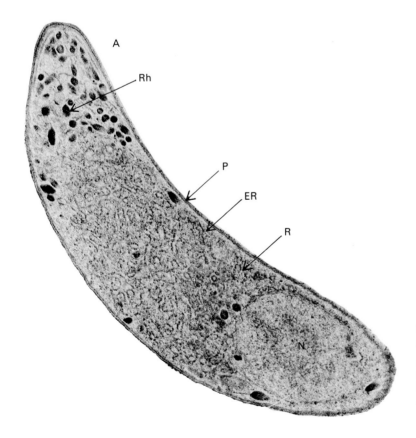

81. An electron micrograph of a section of a mature ookinete.

A number of organelles are shown: the nucleus N, abundant ribosomes R, and endoplasmic reticulum ER, the three-layered pellicle P, the apical complex A, including numerous rhoptries Rh.

Prof. R. E. Sinden

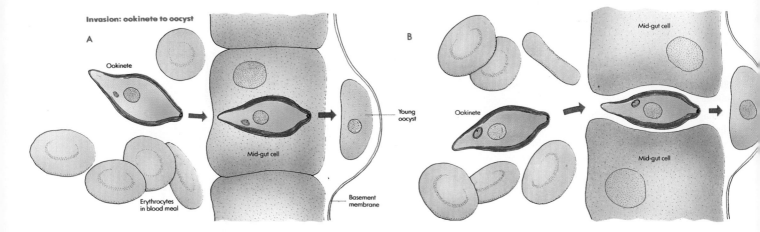

Invasion: ookinete to oocyst

A

Ookinete

Erythrocytes
in blood meal

Mid-gut cell

Young
oocyst

Basement
membrane

B

Mid-gut cell

Ookinete

Mid-gut cell

82. The invasion of a mosquito mid-gut wall.

The ookinete can probably penetrate in alternative ways: through a mid-gut cell (A), or between two mid-gut cells (B). The oocyst develops between the mid-gut cells and their basement membrane.

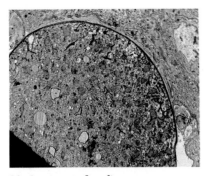

83. An oocyst of medium age.

The crowding of the neighbouring tissues of the mosquito's stomach is caused by the rapid growth of the parasite. The cyst wall is well shown. Peripheral segmentation of the cytoplasm is beginning.   *Prof. R. E. Sinden*

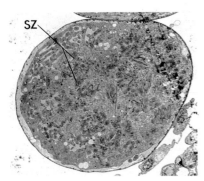

SZ

84. Established sporogony. The cytoplasm is divided into areas of sporoblast separated by vacuolated regions.

Numerous sporozoites SZ are developing into the vacuoles from the apical complex backwards. The cytoplasm appears to have shrunk from the wall of the oocyst.   *Prof. R. E. Sinden*

the ookinete penetrates through or between the epithelial cells (Fig. 82). It comes to rest between the stomach epithelial cells and their basement membrane. Here the apical complex, cytoskeleton, and inner pellicular membranes are resorbed. The organism becomes approximately spherical, and the next phase begins.

3.3   Phase 2. From ookinete to sporozoite Sporogony: the first of the three phases of asexual multiplication

Each asexual phase begins with growth, proceeds through a stage of internal division, and ends as a new generation of invasive parasites is released. The first asexual phase is in the mosquito, and is called sporogony. It is extracellular, unlike the subsequent two asexual phases in man. The accumulation of malaria pigment suggests that the parasite feeds on the haemoglobin of the blood meal, but it is not known if it absorbs nutrients from the mosquito as well. Presumably the mosquito can also cope with the waste products from the parasite, because even heavily infected mosquitoes show few ill-effects.

The oocyst

The embedded ookinete becomes the oocyst, which grows rapidly, and then divides and differentiates internally into sporozoites. The development of the oocyst is the longest active phase in the *Plasmodium* life cycle. It lasts for 8 to 35 days, depending on temperature and the species of *Plasmodium*. The greater the duration of sporogony, the smaller the probability of the mosquito surviving long enough to become infective.

The first changes in the cytoplasm of the oocyst indicate an acceleration of metabolism. The endoplasmic reticulum and Golgi vesicles increase rapidly, while mitochondria proliferate and elongate. These have cristae in all human plasmodia at this stage, indicating oxidative metabolism to provide energy efficiently.

The rapidly growing oocyst is surrounded by a cyst wall (Fig. 83). This is an extracellular coat of fibrous proteins, probably synthesized and secreted by the parasite. As the parasite enlarges the cyst wall is stretched and thinned, finally to burst and release the sporozoites.

Old or degenerate oocyst walls accumulate melanin pigment, and are called Ross's black spores.

The nucleus of the developing oocyst enlarges as the production of thousands of new genetic units begins. It becomes lobed, as invaginations of the nuclear envelope develop. Genetic multiplication continues in the nuclear lobes. Many small mitotic spindles with associated MTOCs form in successive waves on each invagination.

The cytoplasm of the maturing oocyst is now subdivided into large interconnected islands by deep invaginations of the plasma membrane, perhaps associated with dilation of part of the endoplasmic reticulum. The cytoplasm appears to contract, leaving numerous vacuoles containing flocculent material beneath the cyst wall. Lobes of the nucleus enter each cytoplasmic island to form the germinal tissue for sporozoite development, called sporoblast. The apical complex of a sporozoite develops in the cytoplasm and invaginated cell membrane near each nuclear spindle pole and its associated MTOC (Fig. 84).

Differentiation of the sporozoite

Like the ookinete, the sporozoite differentiates from the apical complex backwards, forming an elongated parasite with a three-layered pellicle and a cytoskeleton of longitudinal microtubules. A nuclear lobe moves into each sporozoite bud, followed by a mitochondrion and other cytoplasmic components.

The mature oocyst is a ball of worm-shaped sporozoites dispersed in remnants of sporoblast (Fig. 85). A large oocyst may measure 80 micrometres in diameter, almost visible to the naked eye. From each oocyst of *Plasmodium falciparum* at least 1000 sporozoites eventually develop. Oocysts produce malaria pigment, laid down in patterns distinctive of each species of *Plasmodium*.

When the oocyst bursts sporozoites are released into the body-cavity (haemocoel) of the mosquito. They move anteriorly through the body fluid, penetrate the basal lamina of the salivary glands, pass through the secretory cells, and accumulate in the salivary ducts (Fig. 86). The number of sporozoites in an individual *Anopheles* may be enormous, with the salivary glands so packed with parasites that the ducts appear blocked.

3.4   Phase 3. From sporozoite to hepatic merozoite
Hepatic schizogony: the second phase of asexual multiplication

Liver infection

Sporozoites injected by a feeding *Anopheles* circulate in the blood until they reach the liver (Fig. 87). Here they enter the Kupffer cells. These phagocytic cells are related to macrophages. They line the liver capillaries (sinusoids) and remove bacteria and particulate matter from the blood circulating through the liver. Kupffer cells usually kill and digest bacteria and other micro-organisms, but in some way the sporozoite evades destruction. Indeed, in experiments, sporozoites ingested by peritoneal macrophages survived and destroyed the macrophages which had taken them in. Sporozoites at

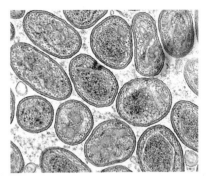

85. Maturing sporozoites, still within the oocyst.

Cross-section of the parasites reveals the three-layered pellicle and the sub-pellicular cytoskeleton of microtubules. Beneath the microtubules the nucleus is seen in cross-section in most of the parasites in this picture.   *Prof. R. E. Sinden*

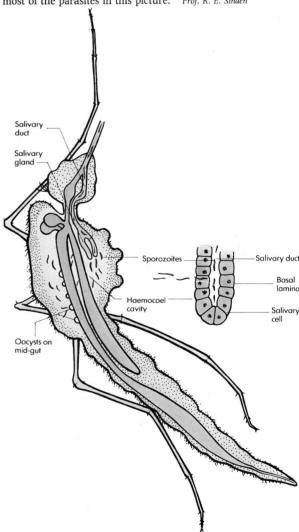

86. The pathway of the sporozoites from ruptured oocyst through the haemolymph in the haemocoel to salivary glands.

The sporozoites penetrate the basal lamina and secretory cells of the gland, eventually to reach the salivary tubules and ducts.

Trustees of the Wellcome Trust

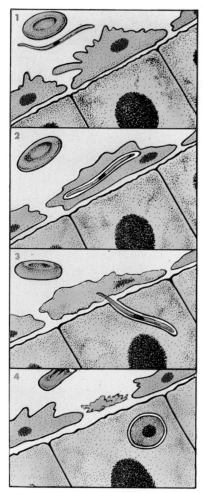

87. A diagram showing the structure of the liver and the process of sporozoite invasion.

1. A sporozoite in the blood among the red blood cells. The blood-vessel is lined by Kupffer cells, which separate the hepatocytes from the blood.
2. A sporozoite has been taken up from the blood and lies in a vacuole within a Kupffer cell.
3. The sporozoite escapes from the Kupffer cell, perhaps destroying it in the process, and invades the underlying hepatocyte.
4. Within the hepatocyte the sporozoite becomes the hepatic trophozoite.

Trustees of the Wellcome Trust

first lie in vacuoles in Kupffer cells, but rapidly move on, to invade the liver cells (hepatocytes).

The hepatic trophozoite

The sporozoite now changes rapidly, losing its apical complex, pellicle, and microtubules, to become the rounded hepatic trophozoite (Fig. 88). Like the oocyst the trophozoite appears to stimulate little host reaction, but degenerating parasitized hepatocytes are found in many preparations, implying that some sort of host defence is activated.

In the warm (38–39°C) nutrient-rich hepatocyte the trophozoite grows rapidly, reaching 40 micrometres in diameter (the size of an oocyst) in only 48 hours, distending and distorting the parasitized cell. In the cytoplasm the endoplasmic reticulum proliferates and mitochondria multiply. The nucleus enlarges and develops lobes. Numerous mitotic spindles appear.

Hepatic schizogony

The next stage is the division of cytoplasm and nucleus, producing 10–30 000 invasive merozoites by the process of hepatic schizogony. Before release of merozoites the multinucleate trophozoite is called a hepatic schizont.

The division of a hepatic trophozoite into merozoites resembles closely the division of the oocyst into sporozoites. Vacuoles of flocculent material develop at the periphery of the trophozoite, and discharge their contents into the parasitophorous vacuole. The cytoplasm is divided by deep clefts from the cell membrane, producing interconnecting islands of cytoplasm (cytomeres). On each cytomere small conical projections are associated with a nuclear lobe. This germinative complex is called meroblast. Apical complexes begin to develop at points where the mitotic spindles in the nucleus bring MTOCs close to invaginated cell membrane. The merozoite develops pellicle and microtubular cytoskeleton from anterior to posterior. A nucleus and a rudimentary mitochondrion move into each merozoite bud (Fig. 89).

88. A young trophozoite developing within a hepatocyte.

The apical complex is still visible on the left of the parasite. Rapid growth is beginning and the apical complex will soon be resorbed. This parasite was found 15 hours after injection of sporozoites.

Dr. J. F. G. M. Meis

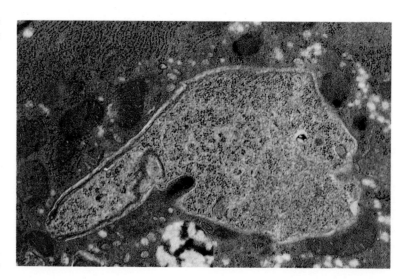

The completed merozoites lie in the parasitophorous vacuole, with residual fragments of meroblast. The hepatocyte now dies and bursts, liberating masses of merozoites embedded in fragments of degenerate cytoplasm into the blood (Fig. 90). The masses are too large to be engulfed by Kupffer cells, and presumably disintegrate after leaving the liver, to free the merozoites.

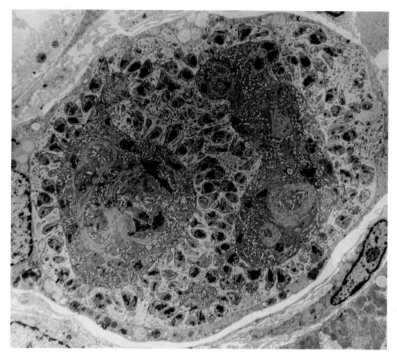

89. A later stage of schizogony.

Numerous meroblasts are budding from the surface of the cytomeres. The cytoplasm of the maturing schizont is becoming a mass of merozoites.

Dr. J. F. G. M. Meis

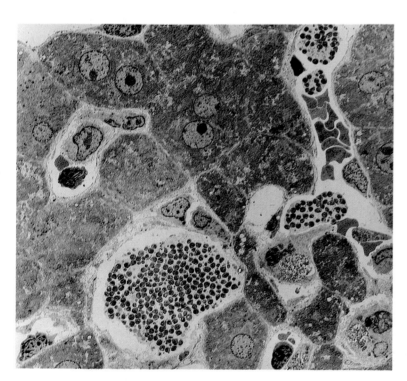

90. Rupture of a mature hepatic schizont, releasing merozoites into a sinusoid.

Clumps of merozoites embedded in degenerate cytoplasm are being released by the rupture of the hepatocyte into a sinusoid. In this form the merozoites evade the Kupffer cells, and are released individually in the peripheral blood.

Dr. J. F. G. M. Meis

91. The first pictures of a hypnozoite.

A section of liver stained by conventional techniques for light microscopy several days after injection of *Plasmodium vivax* sporozoites. There is a late schizont and the arrow identifies a minute hypnozoite. This is a parasite which has entered a resting stage and will not develop for weeks or months.   *Dr. W. A. Krotoski*

The same section stained by the immunofluorescent technique, which causes malarial parasites to shine in ultra-violet light. The large yellow mass is the schizont, and the hypnozoite is shown by the arrow. It was the use of this technique that revealed the presence of hypnozoites in the life cycle of *Plasmodium vivax*.   *Dr. W. A. Krotoski*

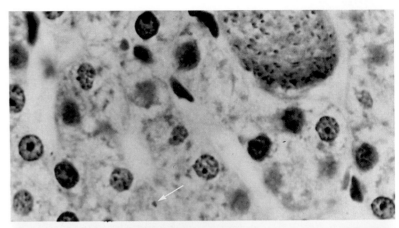

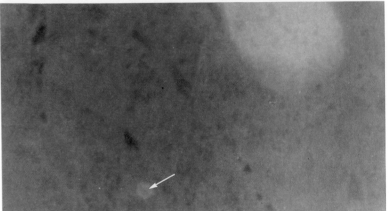

The hypnozoite (Fig. 91)

In *Plasmodium vivax* a proportion of sporozoites entering hepatocytes do not develop immediately. They become rounded and enter a resting phase called the hypnozoite. Weeks or months later development begins and a relapse of illness is initiated. The hypnozoite is very small, and was not discovered until the technique of immunofluorescent staining was applied. The mechanism of delayed relapse in *Plasmodium malariae* infection remains unknown. Hypnozoites have not been found. *Plasmodium falciparum* does not cause delayed relapses.

Cyclical reinfection is an alternative theory to explain delayed relapses. It proposes that some merozoites infect hepatocytes, not erythrocytes, to start new cycles of hepatic schizogony. Many doubted this theory before the hypnozoite was discovered, and it now has few advocates.

3.5   Phase 4. From hepatic merozoite to gametocyte
The third phase of asexual multiplication

In this phase repeated cycles of erythrocytic schizogony produce an enormous multiplication of parasite numbers. For the life cycle to continue a feeding mosquito must ingest at least ten gametocytes; fewer are unlikely to establish infection in the insect. The average mosquito blood meal is 5 microlitres; the average human blood volume is 5 litres; and in a mature infection 10 per cent of the parasites are gametocytes. Thus 100 million parasites in the blood are necessary for mosquito infection. This means that less than 1 per cent of the erythrocytes need to be parasitized for the infection to be transmitted, but in clinical *falciparum* malaria parasitized erythrocytes are often more than 20 per cent of the total.

It is this phase which causes all the illness and death attributed to malaria. This massive multiplication is essential for the survival of *Plasmodium*, and almost all of the millions of new parasites have no future.

It is interesting that there is a parasite of monkeys called *Hepatocystis*, which is closely related to *Plasmodium*, but avoids erythrocytic schizogony. Instead hepatic schizogony is greatly increased, producing large liver cysts which release merozoites for long periods. Each merozoite invades an erythrocyte and matures into a gametocyte. *Hepatocystis* is a much more sophisticated parasite than *Plasmodium*, living in a better relationship with the host, causing little disease, and keeping the host infectious for long periods. Humanity has a good case against *Plasmodium*.

The merozoite (Fig. 92)

The merozoite is the smallest and shortest-lived form in the life cycle. It has the features distinctive of an invasive form—apical complex, pellicle, cytoskeleton—and is specialized to recognize and attach to specific molecules in the membrane of an erythrocyte, and then to invade the erythrocyte itself. The surface molecule recognized varies with the species of *Plasmodium*. In *Plasmodium vivax* it is the Duffy blood-group antigen. In *Plasmodium falciparum* it is probably glycophorin, another glycoprotein of the erythrocyte membrane. The recognition site on the merozoite is at the tip of the apical complex, and it is here that firm initial attachment to the erythrocyte membrane occurs.

Invasion of erythrocytes

A diagram of this process is shown in Fig. 93, and Fig. 94 is a remarkable set of electron micrographs of the progessive stages of erythrocyte invasion. The merozoite is inside an erythrocyte within 20–30 seconds of initial contact and attachment. Merozoite and cell membranes fuse in a ring around the initial attachment site. The ring of fusion then enlarges and moves back over the merozoite,

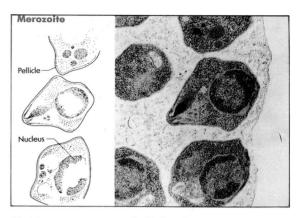

92. Mature merozoites embedded in degenerate cytoplasm in a late hepatic schizont.
Trustees of the Wellcome Trust/Dr. J. F. G. M. Meis

93. Drawings to show the attachment of a merozoite to an erythrocyte, and the subsequent invasion process. *Trustees of the Wellcome Trust*

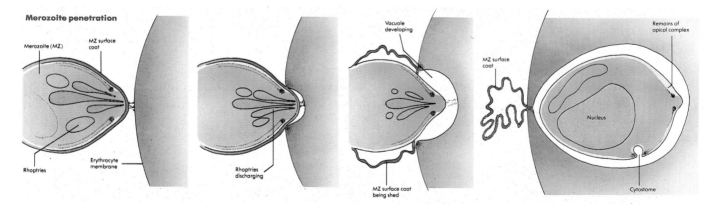

94. A series of electron micrographs of sections of erythrocytes with merozoites in successive stages of invasion.
Prof. M. Aikawa

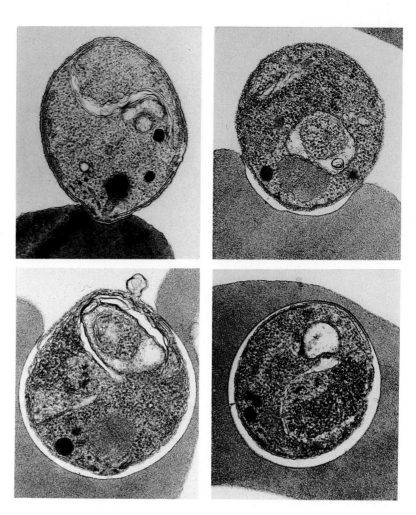

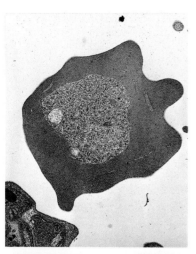

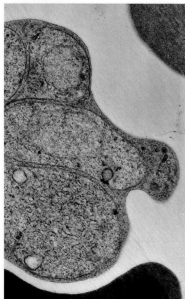

95. Electron micrographs of sections of erythrocytes containing trophozoites.

The cytosome is the point at which erythrocyte cytoplasm containing haemoglobin is taken in to form a food vacuole in the parasite. Trophozoites in erythrocytes often adopt remarkably lobed forms.

Dr. L. H. Bannister
Prof. R. E. Sinden

propelling the body of the parasite into the red cell. The outer surface coat of the merozoite pellicle is shed in the process. The contents of the rhoptries are discharged into the space between the apical complex and the erythrocyte membrane. This appears to alter the properties of the membrane to permit its rapid distension to enclose the entering parasite in a parasitophorous vacuole.

The erythrocytic trophozoite

The merozoite now develops rapidly into the erythrocytic trophozoite (Fig. 95). It loses its apical complex, pellicle, and cytoskeleton. It feeds by ingesting haemoglobin and host-cell cytoplasm through specialized membrane areas called cytostomes. Vacuoles containing ingested food grow within the cytoplasm. The parasite may become remarkably lobed and irregular in shape as it grows. Amoeboid movements can be seen in living parasites, especially in *Plasmodium vivax*. The distinctive ring appearance of the young trophozoites in light microscopical examination is attributed to flattening of the parasite to adapt to the shape of the erythrocyte.

The cytoplasm of the growing erythrocytic trophozoite contains numerous ribosomes, mostly lying free. Endoplasmic reticulum is scanty. Mitochondria are poorly developed at this stage in mammalian parasites, and lack cristae in *Plasmodium vivax* and some other species.

Malaria pigment

As haemoglobin is digested, an iron-haem pigment called haemozoin accumulates in food vacuoles. In human malaria parasites this forms crystals or irregular masses. Chemically malaria pigment is ferric haem, or haematin (Fig. 96). Chloroquine forms molecular complexes with malarial pigment, causing it to form clumps within the cell.

The erythrocytic schizont (Fig. 97)

Schizogony begins after 30–40 hours of growth, depending on the species. Only 8 or 16 merozoites are formed in the very limited space of the erythrocyte. Endoplasmic reticulum proliferates as nuclear division begins. Again the apical complexes of the invasive progeny develop first, where the MTOCs associated with the mitotic spindles approach the plasmalemma. Merozoites develop in a ring at the periphery of the parasite, and clefts do not divide the cytoplasm. As the new merozoites mature from before backwards, pigment masses and degenerating cellular material are sequestered in a residual body, which has pyrogenic properties, contributing to fever induction when the schizont ruptures (Fig.98). Completed merozoites lie free in the parasitophorous vacuole until rupture of the host cell releases them to invade new erythrocytes and repeat the phase 4 cycle (Fig. 99).

Membrane changes

The membrane of the parasitophorous vacuole enlarges during trophozoite development. Its surface area is increased by the development of diverticulae penetrating the erythrocyte cytoplasm. These can be seen by light microscopy and are called Maurer's clefts. The membrane of the parasitophorous vacuole is rich in ATP-ase, and probably has an active role in the transport of nutrients and metabolites to and from the parasite. This membrane originates from the erythrocyte surface membrane, but it appears that the parasite synthesizes and excretes transport enzymes and other proteins, which become inserted into it.

The external erythrocyte membrane is also modified as the parasite matures. Electron-dense material accumulates in patches just beneath the membrane, in association with the formation of knobs or pits. In *Plasmodium falciparum* such knobs are especially prominent, and

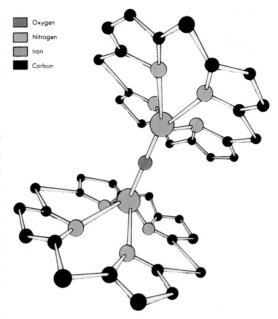

Oxygen
Nitrogen
Iron
Carbon

96. A drawing of the basic sub-unit of the haematin crystal. Two haematin molecules are joined by an oxygen bridge.   *Trustees of the Wellcome Trust*

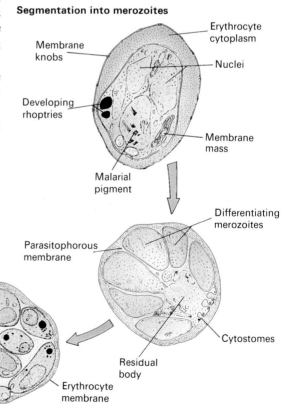

Segmentation into merozoites

Membrane knobs
Erythrocyte cytoplasm
Nuclei
Developing rhoptries
Membrane mass
Malarial pigment
Parasitophorous membrane
Differentiating merozoites
Cytostomes
Residual body
Apical complex
Rhoptries
Pellicle
Erythrocyte membrane

97. Drawings to show the process of schizogony: the segmentation of the trophozoite, producing merozoites.   *Trustees of the Wellcome Trust*

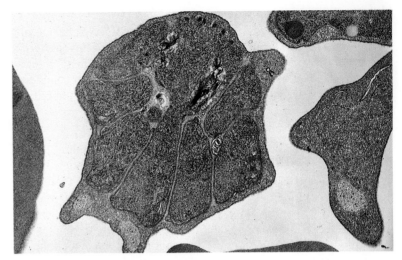

98. Established schizogony: malaria pigment marks the residual body, and several cytostomes persist.

Dr. L. H. Bannister

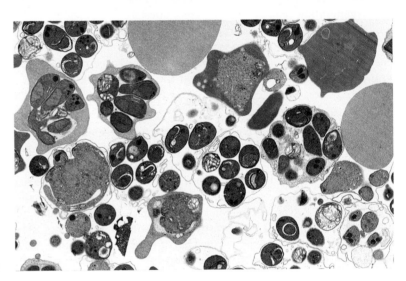

99. A cluster of late schizonts on the point of rupture. *Dr. L. H. Bannister*

may cause parasitized red cells to adhere to endothelial cells in capillaries and venules, an important process called sequestration (see Chapter 6).

Gametocytes

When the cycle of erythrocytic schizogony has repeated several times, a proportion of merozoites develop into gametocytes, not trophozoites, after invading erythrocytes. Gametocytes may be male or female. The stimulus switching the merozoite to this alternative development pathway is not known. Increasing host immunity may be important.

The development of gametocytes has been studied most in *Plasmodium falciparum*, because this species is the easiest *Plasmodium* infecting humans to grow in culture, and the ten- or eleven-day maturation of its gametocytes permits detailed analysis of the process. However, remember that *Plasmodium falciparum* is in some respects not a typical mammalian species, especially in the features of its gametocytes.

About 24 hours after red-cell invasion of the host erythrocyte the developing gametocyte begins to form a new three-layered pellicle and a cytoskeleton of microtubules. It becomes ovoid, but may be distorted by uneven development into a crescent or D shape. Further growth of cytoplasm and microtubules leads to the full-size gametocyte, held to a cigar-shape by its cytoskeleton and stretching the erythrocyte membrane. Finally the microtubules disappear and the tensions in the membrane pull the now less rigid gametocyte into the crescentic form, from which the species derives its name (Fig. 100).

In the female gametocyte nucleus a mitotic spindle is formed, but mitosis progresses no further until a male genome enters at fertilization. The genome of the male nucleus is replicated three times. Thus the male cell nucleus is octoploid and larger than the female nucleus. The female gametocyte maintains numerous ribosomes and mitochondria, and a nucleolus is prominent. In the male ribosomes and mitochondria are reduced (Fig. 101).

Small dense osmiophilic bodies resembling micronemes can be seen in the periphery of gametocytes of both sexes. These are discharged when the cell is activated, and probably disrupt the host-cell membrane as the gamete is released. Malarial pigment granules are found around the nucleus.

Mature gametocytes have reduced metabolism. DNA replication has stopped, and there is little RNA or protein synthesis. In consequence gametocytes are not susceptible to most antimalarial drugs.

Gametocyte structure

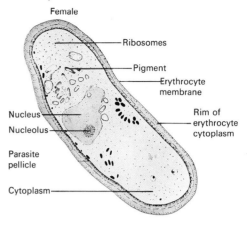

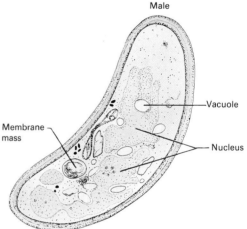

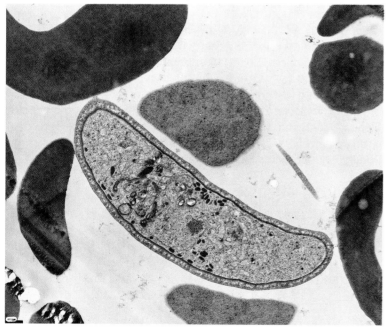

101. Mature gametocytes of *Plasmodium falciparum*.

A mature male gametocyte of *Plasmodium falciparum*, showing its large nucleus and poorly developed ribosomes and endoplasmic reticulum. The parasite membrane has three layers. It is still enclosed by the erythrocyte membrane.

Prof. R. E. Sinden

A mature female gametocyte of *Plasmodium falciparum*. It has a smaller nucleus than the male, with a nucleolus. Ribosomes are more frequent in the cytoplasm. There are three layers in the parasite's external membrane.

Prof. R. E. Sinden

100. Male and female gametocytes of *Plasmodium falciparum*.

In the male the nucleus predominates: it contains 8 genomes. In the female the cytoplasm is more abundant and appears more active.

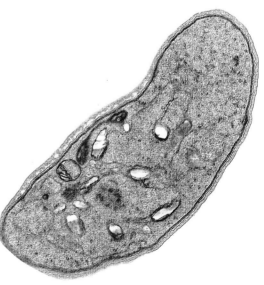

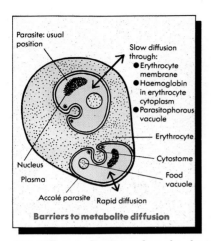

102. Diffusion of nutrients from the plasma to the parasite is faster in the 'accolé' position, often adopted by *Plasmodium falciparum*.

3.6 Metabolic pathways in *Plasmodium*

The metabolism of the erythrocytic trophozoite is known in some detail. At this stage the parasites can be harvested in reasonable quantities from cultures, removed from the erythrocyte, and separated from other cells. In such preparations the metabolic activity of the parasite can be distinguished from the metabolism of erythrocytes and white cells. Very little is known about the metabolism of other stages in the life cycle. More information is badly needed, especially about points where the metabolic activities of the parasite and host cell differ. These are potential targets for selective drugs, able to injure the parasite while leaving host cells unharmed.

The parasite is separated from the plasma by the erythrocyte membrane, a layer of haemoglobin, and the membrane of the parasitophorous vacuole. These barriers must slow diffusion of nutrients into the parasite and waste products out, limiting the rate of growth. Diffusion can be increased in at least two ways. The parasite may apply itself to the erythrocyte membrane, as in the 'accolé' forms of *Plasmodium falciparum*, or make amoeboid movements, as in *Plasmodium vivax* (Fig. 102).

Figure 103 is a plan of the main metabolic pathways known in *Plasmodium*.

Energy supply

Glucose fuels *Plasmodium* metabolism in the erythrocyte. It is used in the glycolytic pathway to generate energy as adenosine triphosphate (ATP), producing lactic acid as an end-product. This pathway is anaerobic—no oxygen is used. As the erythrocyte circulates its haemoglobin is alternately oxygenated and deoxygenated, but the parasite is not affected by these fluctuations. Glycolysis with anaerobic energy-generation is inefficient, and a lot of glucose is taken from the blood. In heavy infections the patient's blood-glucose concentration may fall dangerously (hypoglycaemia), a process which may be exacerbated during treatment, because quinine stimulates insulin secretion.

Protein synthesis

Amino acids for protein synthesis come mostly from digested haemoglobin. Each molecule of haemoglobin digested releases four molecules of the red iron-containing pigment haem. This accumulates in the food vacuoles, becomes oxidized to the brown ferric haem or haematin, and precipitates or crystallizes as malarial pigment (Fig. 104).

Nucleic acids

Important components of nucleic acids are the two purines, adenine and guanine, and the pyrimidines cytosine and thymine (in DNA) or uracil (in RNA). *Plasmodium* takes its purines from the host plasma, but synthesizes its own pyrimidines. This idiosyncracy remains unexplained and unexploited. The nucleoprotein of haploid *Plasmodium* has some special properties. It is deeply stained by Azure pigments derived from oxidized methylene blue, the basis of the Giemsa and other staining procedures. It interacts with chloroquine, quinine, and similar drugs, which appear to slide into the stacked

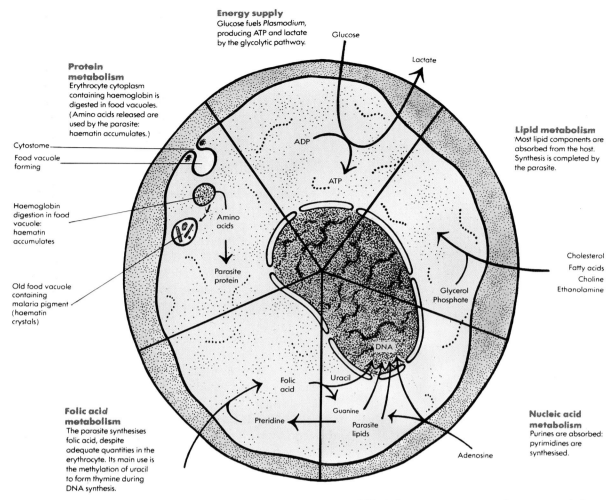

Energy supply
Glucose fuels *Plasmodium*, producing ATP and lactate by the glycolytic pathway.

Glucose

Lactate

Protein metabolism
Erythrocyte cytoplasm containing haemoglobin is digested in food vacuoles. (Amino acids released are used by the parasite: haematin accumulates.)

Lipid metabolism
Most lipid components are absorbed from the host. Synthesis is completed by the parasite.

Cytostome

Food vacuole forming

ADP

ATP

Amino acids

Cholesterol
Fatty acids
Choline
Ethanolamine

Haemoglobin digestion in food vacuole: haematin accumulates

Parasite protein

Old food vacuole containing malaria pigment (haematin crystals)

Glycerol Phosphate

DNA

Folic acid metabolism
The parasite synthesises folic acid, despite adequate quantities in the erythrocyte. Its main use is the methylation of uracil to form thymine during DNA synthesis.

Folic acid

Uracil

Guanine

Pteridine

Parasite lipids

Adenosine

Nucleic acid metabolism
Purines are absorbed: pyrimidines are synthesised.

103. A diagram to summarize the main metabolic pathways in *Plasmodium*.

purines and pyrimidines of nucleic acids, possibly interfering with replication and transcription.

Lipids

Lipid synthesis is especially important in the production of the various membrane systems of the parasite. The subject is complicated, but it is likely that the parasite synthesizes some lipid components, and takes up others from the host. It can synthesize glycerol, the basis of triglycerides and phospholipids, and it can phophorylate lipid precursors. From the host it requires fatty acids, cholesterol, and probably ethanolamine and choline.

Folic acid

Folic acid is a vitamin: a cofactor essential for a number of important reactions, including the methylation of uracil to form thymine during DNA synthesis. It has a complicated metabolic cycle in most organisms, requiring reactivation after each use. *Plasmodium* synthesizes its own folic acid, although ample supplies are available in the plasma and in the erythrocyte. An essential precursor is para-aminobenzoic acid (PABA). Sulphonamide antimicrobial drugs

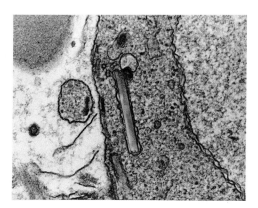

104. A crystal of haematin within the cytoplasm of a trophozoite.
Prof. R. E. Sinden

interfere with PABA metabolism, and some have antimalarial activity. Milk is deficient in PABA, and a milk-only diet can protect against malaria.

Dihydrofolate reductase is an enzyme essential in folic acid activation. It is inhibited by two useful antimalarial drugs, proguanil and pyrimethamine. The equivalent human enzyme is inhibited only at much greater drug concentrations.

3.7 Molecular biology of *Plasmodium falciparum*

The genetic structure of *Plasmodium* can now be studied using the powerful analytical methods of molecular biology. No detailed description is yet possible, but research proceeds rapidly.

Each nucleus of *Plasmodium falciparum* contains about 0.02 picograms of DNA: that is 2×10^{-14} grams, or two hundred million millionths of a gram. This corresponds to a total of approximately ten million base-pairs: about ten times more than the genome of a typical bacterium, and about one tenth of the genome of simple worm.

The adenine-thymine base-pair (A + T) is more than 80 per cent of the bases in *Plasmodium falciparum* DNA. This proportion is high in the other *Plasmodium* species studied, but in *Plasmodium falciparum* it is the highest yet recorded for a living organism. A consequence of the unusual (A + T) ratio is that the genetic code is more restricted—less degenerate—in *Plasmodium falciparum* than it is in other organisms. Two other observations may be linked to the high (A + T) ratio.

- *Plasmodium* trophozoites depend on the host plasma for their supply of purines, and in man adenine in the form of adenosine is the predominant plasma purine.
- *Plasmodium* trophozoites synthesize their own thymine, in a reaction requiring folic acid. The unusually high requirement for thymine may explain the sensitivity of the organism to antifolate drugs and PABA deficiency.

Genetic relations of *Plasmodium falciparum*

Plasmodium falciparum is genetically more closely related to malaria parasites from rats and birds than it is to other human *Plasmodium* species (Fig. 105). Its adaptation to man may have been relatively recent, and perhaps is still incomplete. This conclusion is supported by studies of the (Guanine + Cytosine) values for the DNA of different species of *Plasmodium*, and by the pattern of interaction of the DNA with probes for specific genes. Three types of *Plasmodium* species can be distinguished by their (Guanine + Cytosine) values: N.B.: (G + C) = 1-(A + T).

- Those with a single component genome containing about 18 per cent (Guanine + Cytosine) : e.g. *Plasmodium falciparum* (human).
- Those with a single component genome containing about 30 per cent (Guanine + Cytosine) : e.g. *Plasmodium knowlesi* (monkey).
- Those with two or more genome components, one containing about 18 per cent, the other about 30 per cent, (Guanine + Cytosine) e.g. *Plasmodium vivax* (human).

Species	Host	(G + C) content		gene hybridization			
		18%	30%	Actin	Tubulin	DHFR	TS
P. falciparum	Man	+	–	–	–	–	+
P. berghei	Rodent	+	–	–	–	–	+
P. lophurae	Bird	+	–	–	–	–	+
P. knowlesi	Monkey	–	+	+	+	+	–
P. fragile	Monkey	–	+	+	+	+	–
P. cynomolgi	Monkey	+	+	+	+	+	–
P. vivax	Man	+	+	+	+	?	?

Note: P. vivax and P. cynomolgi each have two components in the genome, which differ in (A + T) content.

Actin = chicken actin gene.

Tubulin = Chlamydomonas tubulin gene.

DHFR = mouse dehydrofolate reductase gene.

TS = yeast thymidylate synthetase gene.

105. The relationships of different *Plasmodium* species, assessed by the (G + C) content of the genome, and its ability to bind gene probes.

Plasmodium falciparum is more closely related to bird and rodent *Plasmodium* species than it is to other human and primate species.

'Chromosomes' of *Plasmodium falciparum*

Chromosomes cannot be distinguished by light or electron micrographs of *Plasmodium falciparum* cells in division. Chromosome-like individual molecules of DNA can be separated by modern methods of electrophoresis. Individual genes can be allotted to DNA molecules identified, and genetic maps made. In total there may be ten or more such 'chromosomes' in each cell of the parasite. The size and characteristics of identified DNA molecules are constant in the asexual and sexual blood stages of a single strain, but vary between strains isolated from different geographical areas, or from different clones isolated from the same patient. Variability in length indicates that the genetic material may undergo rearrangements such as gene duplication or deletion from time to time. Parasites kept in culture for many generations frequently lose important genes, perhaps by deletion during such rearrangements.

3.8   *Plasmodium* antigens

These are potential targets for malarial vaccines.

Sporozoites and merozoites are the only forms of *Plasmodium* which are exposed to serum antibodies. All other forms in the human are intracellular, and so shielded from this expression of immune defence. Gametes released when gametocytes are activated in the stomach of a mosquito are exposed to antibodies in the blood meal, but usually specific antibodies are absent (gametocytes are intracellular while circulating in the blood). In experiments antibodies to gamete proteins block transmission to mosquitoes.

Surface proteins of sporozoites and merozoites may be useful as malaria vaccines, but experience so far is discouraging. Immunization using gamete products raises the interesting prospect of an 'altruist's vaccine', an immunization accepted for no benefit to the individual, but to prevent transmission and so protect others.

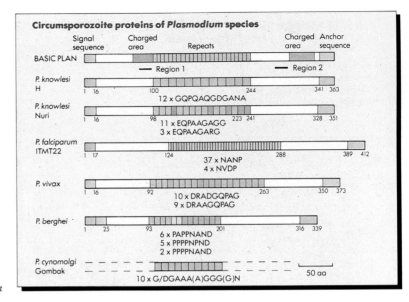

106. Circumsporozoite proteins of *Plasmodium* species.

Repeating sequences of amino acids are a frequent and remarkable feature of *Plasmodium* antigens.

107. The circumsporozoite protein (CSP) reaction.

If antibodies bind to the circumsporozoite protein the complexes separate and are shed from the rear of the parasite.

Sporozoite proteins

The circumsporozoite protein (CSP) is the dominant protein on the surface of mature sporozoites. Each molecule is a single polypeptide chain outside the sporozoite, anchored at one end to the pellicle. Its most remarkable feature is the centre section of the polypeptide, which consists of a short sequence of amino acids, repeated many times (Fig. 106). The composition, length, and number of repeat sequences vary between species of *Plasmodium*, and between strains of the same species.

Thus a common CSP from *Plasmodium falciparum* contains 37 repeats of the tetrapeptide asparagine-alanine-asparagine-proline, with 4 interspersed copies of a variant (asparagine-valine-aspartic acid-proline). Each repeat is antigenic.

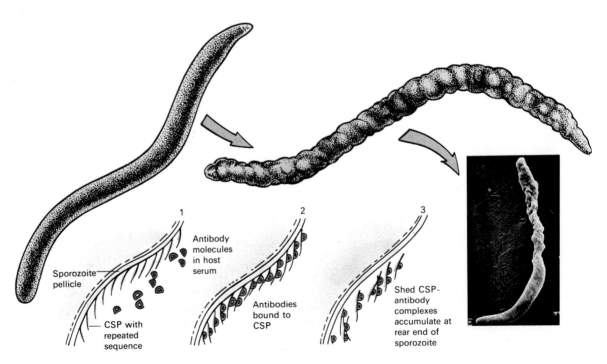

If host antibodies bind to CSP on a sporozoite, the CSP-antibody complexes are shed as a precipitate which collects at the rear of the parasite (Fig. 107). This is called the circumsporozoite protein reaction, and may be a form of defence against host immunity.

Other *Plasmodium* antigens

A number of antigenic proteins identified in parasites, erythrocytes, and plasma during malaria infection are listed in Fig. 108. Some may stimulate immunity-giving partial protection, and may be potential vaccines.

Other antigens stimulate antibody production, but appear to give no protection. They may be part of an 'immunological smokescreen' of variable and strongly antigenic materials released by the parasite

108. Antigenic proteins found in erythrocytes containing trophozoites and schizonts of *Plasmodium falciparum*.

Title	Full name	Location	Molecular weight (thousands)	Comments
S-Antigen	Serum antigen	Parasitophorous vacuole		Very variable antigenically
GBP	Glycophorin-binding protein	Merozoite surface & parasitophorous vacuole	105-115	Status uncertain
PMMSA	Precursor to the major merozoite surface antigens	Schizont & merozoite surfaces	180-220	Possible vaccine protein
RESA	Ring-infected erythrocyte surface antigen	Membranes of erythrocytes containing early trophozoites. Merozoite micronemes	155	Probably injected into the erythrocyte membrane during merozoite invasion. Possible vaccine
KAHRP	Knob-associated histidine-rich protein	Knobs developed on erythrocyte membranes as trophozoites mature	85-105	Implicated in the sequestration of erythrocytes containing schizonts
MESA	Mature parasite-infected erythrocyte surface antigen	Cytoskeleton of infected erythrocytes	about 250	Variable in size and antigenicity
FIRA	*Falciparum* interspersed repeat antigen	Within infected erythrocytes outside the parasites	more than 300	**Strongly antigenic but not protective**: a 'smokescreen' **molecule (see text)**

Table title: **Asexual blood stage antigens of *P. falciparum***

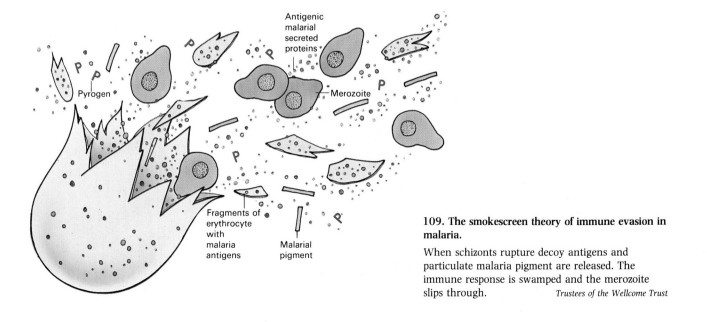

109. The smokescreen theory of immune evasion in malaria.

When schizonts rupture decoy antigens and particulate malaria pigment are released. The immune response is swamped and the merozoite slips through.   *Trustees of the Wellcome Trust*

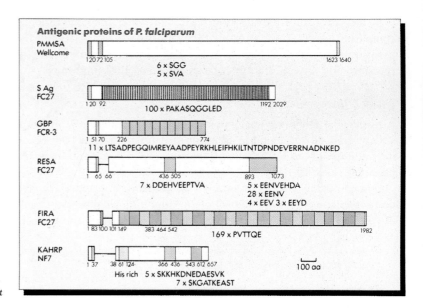

110. Antigenic proteins of *Plasmodium falciparum*.
Trustees of the Wellcome Trust

to swamp and blind the immune defence systems (Fig. 109). The smokescreen is augmented by particulate matter released at the same time: malaria pigment crystals, and erythrocyte fragments. These blockade the body's macrophage defences, and so further impair the development of immunity. A smokescreen must be dense to be effective, which may mean there is a survival advantage to *Plasmodium* in synchronized maturation and rupture of schizonts. The intermittent fevers of malaria might be explained in this way.

Each antigenic protein contains repeat sequences, developed in some to an extraordinary degree (look at S-antigen and FIRA in Fig. 110). Repetitive sequences are also present in the *Plasmodium* genes coding for these proteins. Repeat sequences confer many antigenic sites on each protein, and may permit rapid variations in protein size and sequence. These may be important properties if immune evasion by the 'smokescreen method' is indeed their function.

4 Malaria in the community

Epidemiology is the study of the patterns of disease in human communities, and of the various factors which determine when diseases occur, and how they spread. It is the science basic to public health and disease control. Two words need definition.

- The *incidence* of a disease is the number of new cases occurring in a population during a stated period.
- The *prevalence* of a disease is the number of cases in a population at a stated time: this number includes new and continuing cases.

4.1 Stable and unstable malaria

Malaria in the individual varies from a severe or fatal illness to symptomless circulation of parasites in the blood. Similarly, malaria in the community varies from a devastating epidemic to an insidious continual problem. The factors controlling the pattern, amount, and severity of malaria in the community must be studied if we are to control malaria and to understand the different presentations of malaria throughout the world. The contrasts may be great.

A doctor from the Punjab or Sri Lanka might report thus (Fig. 111).

Malaria is a great problem in my area. It is like influenza in England. We may go for years with very little malaria, and then suddenly there is a great epidemic. Everyone is ill. The trains and buses stop running; there is a food shortage in the cities as the farmers and the truck drivers are ill: everyone is affected, men and women, adults and children. Few die at first, but more die later if the epidemic is a long one. And then it all goes away for some years.

From Cameroon in West Africa the doctor's report is different (Fig. 112).

Malaria is a great problem in my area too. But we never see epidemics. Malaria is around all the time. Everyone gets it: by the time children are two years old most have malaria, and they go on getting it. If you prick the fingers of 100 schoolchildren, 75 or more of them will have malaria parasites in their blood on any day. We think about 5 per cent of all children born die of malaria. The survivors get ill less often as they get older, and adults have only occasional fevers.

In the Punjab, Sri Lanka, and many other places malaria is called unstable, because of the epidemics and the way it comes and goes. In much of Africa, Papua New Guinea, and elsewhere, the regular presence of malaria is called stable malaria. In some parts of the Sahel (the seasonally dry area just south of the Sahara desert) there is a seasonal variation in malaria, but it is the same each year, and so it is still called stable malaria.

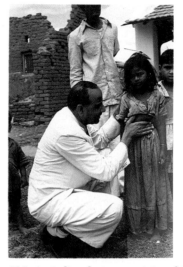

111. An Indian doctor examining children for enlargement of the spleen. *Prof. L. J. Bruce-Chwatt*

112. Collecting a blood sample from a child ill with fever in Cameroon. *Prof. L. J. Bruce-Chwatt*

In unstable malaria, epidemics affect all ages of people, and they are not immune. In stable malarious areas the infection is most severe in children. Adults have acquired some immunity, and parasites are not found so often in their blood, even though they are still bitten by infective mosquitoes. Young babies have some immunity from their mothers.

4.2   Transmission of malaria

In some ways, malaria is a simple disease—it is either in people or in mosquitoes. It cannot hide anywhere else. Transmission depends on three things: infected people, infected mosquitoes, and the biology of the parasite in both its hosts.

Infected people

The best way to measure how malaria can spread is to ask how many new cases of malaria occur, on average, from one person with malaria, before he or she dies or gets better. This is called R or the 'basic case-reproduction rate'. So if 1 person with malaria leads to mosquitoes being infected and passing on the infection to 3 other people, R is 3. Where malaria is stable, R cannot be measured directly: everyone will already have the disease or be immune.

R is an interesting way to measure malaria (Fig. 113). If it is above 1, malaria will spread, because each case will cause more than one new case of malaria until everyone has been infected. If it is much above 1 then malaria will spread quickly: if it is only just above 1 it will spread slowly. If R goes below 1, then malaria will die out, because fewer people are infected than before each time round the cycle. R is not constant: it will be much greater than 1 as an epidemic begins, and fall below 1 as the epidemic dies away. To control malaria, reduce R below 1.

Infected mosquitoes

How do mosquitoes affect the spread of malaria? Much is known about Anopheline mosquitoes, but only four aspects of mosquito life directly affect malaria transmission.

1. Mosquito density.
How many mosquitoes are there? R is proportional to mosquito density.

$$R\alpha \text{ Density}$$

The more mosquitoes, the more malaria spreads. This is not surprising, and the relation is direct proportionality: if you treble the number of mosquitoes, you treble R.

2. Mosquito feeding.
How often do the mosquitoes feed? What proportion of meals are on people (and not other animals)?

$$R\alpha \text{ (chance of feeding on man).}$$

The mosquito has to feed on people to catch malaria. Later it must feed on people again to pass it on. So R increases the more the mosquito feeds on man.

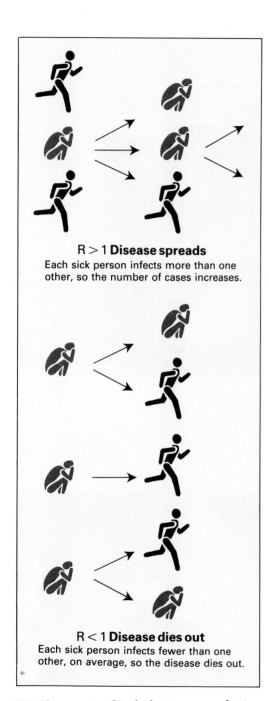

R > 1 Disease spreads
Each sick person infects more than one other, so the number of cases increases.

R < 1 Disease dies out
Each sick person infects fewer than one other, on average, so the disease dies out.

113. The meaning of R, the basic case-reproduction rate.

3. Mosquito survival.
How long do the mosquitoes live?

$$R \alpha \; p^n$$

That is, R is proportional to the chance of mosquito survival through one day (p), to the power of (n): the number of days needed by *Plasmodium* to complete its development in the mosquito. The mosquito has to feed to catch malaria. Then it has to survive long enough for the parasite to develop and mature to infect other people, five feeding/egg-laying cycles at least, a long time in a mosquito's life. Only elderly mosquitoes can pass on malaria, and many mosquitoes never reach old age in the wild. This is where temperature matters: the warmer the weather, the faster malaria can develop in the poikilothermic mosquito.

Mosquito survival is the single most important factor in malaria transmission.

4. Mosquito susceptibility.
Can the malaria parasite survive in the mosquito? No, if it is in a *Culex* mosquito. Usually yes, if it is in a local *Anopheles* mosquito, but sometimes malaria from one place cannot survive in any mosquito from somewhere else. This is important for reducing the chance of British mosquitoes spreading imported malaria parasites, but it is not important in endemic areas.

4.3   The malaria survey

The objectives of a malaria survey are to evaluate the amount of malaria in a community or district, and to estimate the degree of transmission. The *spleen rate* is a good method of rapid assessment. This is the proportion of the population which has palpable enlargement of the spleen (Fig. 114). It is usually determined for a sample of children aged two to eight years, using a reliable technique for spleen palpation. The proportion of adults with enlarged spleens can be measured similarly. The spleen rate can be refined if the degree of enlargement is also recorded. Although this increases complexity and takes extra time, it does allow calculation of the *average enlarged spleen index* (AES), which is more sensitive to changes in malaria transmission than the simple spleen rate.

The *parasite rate* is the other important index of the prevalence of malaria in a community. This is the proportion of the population with parasites in the blood. (In endemic areas this may be the majority of apparently healthy people.) To be valuable the parasite rate must be measured for a number of narrowly defined age-groups, such as toddlers, small children, school-children, adolescents, and adults.

From such measurements the magnitude of the stress of malaria on a given population can be classified, according to the scheme adopted by the World Health Organisation.

1. Hypoendemic malaria. Spleen rate in children less than 10 per cent. Transmission is weak and the general effects of malaria on the population are small.
2. Mesoendemic malaria. Spleen rate in children between 10 and 50 per cent. This is typical of rural villages in sub-tropical zones.

114. Enlargement of the spleen caused by chronic malaria in a group of Nigerian children.

(N.B. This is not tropical splenomegaly syndrome, which is a disease of young adults).

Royal Tropical Institute, the Netherlands

3. Hyperendemic malaria. Spleen rates in children always more than 50 per cent, and in adults over 25 per cent. Transmission is intense but seasonal, but immunity in the population is insufficient to prevent all age groups suffering symptomatic malaria.

4. Holoendemic malaria. Spleen rate in children greater than 75 per cent, but low in adults. Transmission is intense and continuous, with the development of premunition in adults. A high prevalence is expected of Burkitt's lymphoma, other complications of chronic malaria, and genetic blood disorders

5 Control of mosquitoes and malaria

Malaria control was practised long before the parasite was discovered. The Greeks and Romans in classical times recommended that houses should be built in dry, elevated positions to avoid fever, and the health benefits of draining marshes were known. Malaria epidemiology became a science with the discovery of the role of mosquitoes in malaria transmission by Ross in 1897. Ross immediately recognized the practical implications of his discovery for the control and prevention of malaria. He went on to make practical studies of malaria control, and to begin the mathematical analysis of malaria transmission.

We now know that the ecology of *Plasmodium*, *Anopheles*, and man is complicated and variable. It is difficult accurately to predict the results of any proposed intervention. Usually a programme of house-spraying will reduce the number of malaria cases, but sometimes the effect is small, and on occasion the number may even increase. Fortunately our increasing understanding makes prediction more reliable.

While admitting the difficulties and pitfalls of malaria control, the words of G. W. Hackett in 1937 are still valid, and perhaps more relevant.

The mechanism of malaria transmission is so complicated and delicate that it has never been able to resist any long-continued sabotage. Persistence is more important than perfection and whether control is a partial failure or a partial success depends on the point of view. Above all, let us not allow ourselves to be discouraged by theorists . . . and fight the disease now with weapons already proved useful, albeit imperfect, rather than to fold the hands while awaiting a problematical *therapia magna* of the future.

The strength of malaria is in the enormous reproductive capacity of parasite and mosquito: its weakness is the long development-time before an infected mosquito becomes infectious to man.

5.1 Methods of malaria control

The malaria transmission cycle has three components: the malaria parasite, man, and the *Anopheles* mosquito vector. Control measures can be directed at each (Fig. 115).

The malaria parasite

Complete cure of clinical malaria requires several days' treatment with several drugs. A single-dose curative treatment still eludes the pharmacists. This creates problems of cost and administration.

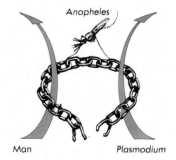

115. The life cycles of man and *Plasmodium* are linked by the life cycle of *Anopheles*.

Breaking the link separates man and *Plasmodium*, and *Plasmodium* will perish.

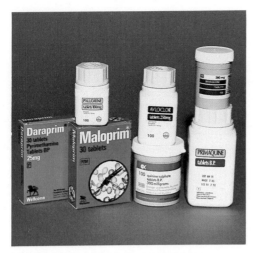

116. Antimalarial drugs in common use.

117. Public health education in Papua New Guinea.
Prof. L. J. Bruce-Chwatt

118. Public health education at Ho in Ghana.
Prof. L. J. Bruce-Chwatt

Prophylaxis using drugs has been of great benefit and widely used. Mass administration of quinine assisted early malaria-control projects in Italy and other countries. Quinine taken in gin to disguise the bitter taste was the original 'gin and tonic' used by the British in India. Chloroquine tablets have a bitter taste, but are safe and have to be taken only once a week. Proguanil is also safe, but has to be taken daily (Fig. 116).

Unfortunately drug prophylaxis is no longer effective in many tropical areas, because resistance is wide-spread and increasing. Vaccines are in active development, but the prospects for a single-course vaccine giving enduring immunity are poor.

Man

Like the mosquito, man is a vector of the parasite. In endemic regions carriers can be identified and treated. Individuals and groups can be educated about precautions to avoid infection. Thus in some circumstances bed-nets can give valuable protection.

The mosquito

Mosquito control is in practice the measure which gives the greatest benefit in the shortest time to the largest number of individuals in a community. Mosquitoes can be attacked in several ways, but the most effective is to spray the interior walls and ceilings of houses with a persistent insecticide (Fig. 119). This method is most likely to kill female Anophelines, as they rest after blood meals. It was the basis of the global efforts to eradicate malaria in former years, and was dramatically successful at first. Unfortunately mosquito resistance to insecticides now limits the value of this technique.

Measures to control mosquitoes can be directed against larvae, living in water, or against adults. Control of larvae is favoured where breeding places are discrete, and identifiable, especially where water collections are few relative to the number of houses. Larval control is therefore particularly useful in towns. A problem with larval control is that any breeding places missed can give rise to numerous adults within a short time.

Adult control aims to reduce the life expectancy of female *Anopheles* until it is less than the duration of sporogony at the prevailing temperatures. The mosquito dies before she becomes infectious. In average tropical conditions this requires a mean life expectancy of less than two weeks for each female *Anopheles*.

Insecticides

Insecticides may be used as space sprays (Fig. 120), or as applications to a surface on which mosquitoes may alight. Space sprays require large volumes, but the insecticide can be short-acting. Surface sprays need smaller volumes, but the insecticide should be persistent, colourless, and odourless. The types of insecticide available are shown in Fig. 121.

The value of insecticides in malaria control is much reduced. Some 50 *Anopheles* species are now resistant to one or more insecticides. Specifically:

119. Spraying beneath the roof: a favourite resting place for mosquitoes. *World Health Organization*

120. Space-spraying of insecticide using the 'Swingfog' machine. *Prof. L. J. Bruce-Chwatt*

Insecticides useful in mosquito control

- For house-spraying against adult mosquitoes:
  - Non-residual (used as space sprays) *Pyrethrum*
  - Residuals (used to spray surfaces) Organochlorines, e.g. *DDT* Organophosphates, e.g. *Malathion* Carbamates, e.g. *Propoxur* Synthetic pyrethroids, e.g. *Permethrin*

- For larval control
  Juvenile hormones or mimics e.g. *Methoprene* Chitin inhibitors, e.g. *Diflubenzuron* Some residuals

Insecticides can be used in a variety of formulations:

- Indoor use
  Where surface marking is unacceptable solutions and emulsions are used. Wettable powders, water dispersible are most widely used. Usually delivered by a portable pneumatic spray machine.

- Outdoor use
  Dusts
  Aerosols
  ULV (ultra low volume) sprays
  Smoke generators
  High-spreading oil larvicides
  Often require specialized application machinery.

121. Table of insecticides. *Trustees of the Wellcome Trust*

122. Aerial spraying of insecticide in the West Indies. *Mosquito Research & Control Unit, Cayman Islands*

- 49 species are resistant to DDT,
- 24 species are resistant to organophosphates,
- 14 species are resistant to carbamates,
- 10 species are resistant to pyrethroids, and
- 14 species are resistant to three or more insecticides.

At least 11 of the 50 resistant species are important malaria carriers. Resistance can be overcome for a time by the use of new or second-choice insecticides, but these are likely to be more expensive, more toxic, and less persistent.

Agricultural use of insecticides is a common cause of resistance, especially when applied by air (Fig. 122). Mosquito breeding areas

123. Fish such as guppies eat mosquito larvae.

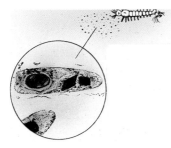

124. An electron micrograph of a section of *Bacillus thuringiensis*, showing two crystals of toxin and a spore.

If a mosquito larva eats the bacillus the toxin paralyses its stomach and makes the contents acid. The spore will then germinate and the resulting bacilli infect the insect and kill it.

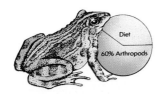

125. Intensive hunting of frogs for export of frogs' legs has caused an increase in malaria in some parts of India.

are usually at the edges of cultivated ground, and so escape a full dose of insecticide, giving good conditions for resistance to develop.

House occupiers commonly welcome the first spraying of their homes. Later they resent the inconvenience and perhaps the odour of repeated sprays. Spraying of temples, mosques, and other sacred buildings may be prohibited. Refusals lead to reduced overall coverage and impair mosquito control. They are a type of resistance in the human population, and may be just as much a problem as resistance in the insect.

Environmental control

This includes all methods of reducing the number of breeding sites for mosquitoes, such as land drainage. It is by far the oldest method in use. Not all collections of water breed vector mosquitoes. Careful species identification and location are necessary if resources are to be used efficiently. Health, public works, and agricultural authorities must work together.

Environmental control is often very expensive. It may be practicable only near main towns. However, local small communities can often achieve much by filling in small ponds, and keeping the edges of streams and irrigation canals free of vegetation.

Biological control

Various types of biological control methods have had some success. Fish such as guppies (Fig. 123), which breed rapidly and eat mosquito larvae, may be valuable, especially in water-tanks and other enclosed water collections. Two bacteria which produce toxins and kill larvae which eat them are *Bacillus thuringiensis* and *Bacillus sphaericus* (Fig. 124). Numerous other mosquito pathogens and predators have been tried: viruses, bacteria, protozoa, fungi, and nematode worms.

Male mosquitoes can be sterilized by chemicals or X-rays, but will still mate. Release of large numbers of infertile males can reduce mosquito breeding, because the female *Anopheles* mates once only in her lifetime.

An example of failure of biological control comes from India. The ruthless hunting of frogs for commercial export of frogs' legs to Europe has reduced the native frog population severely in some areas (Fig. 125). This has allowed mosquitoes and other insects to multiply with little predation. Malaria has increased sharply in consequence.

Personal protection

Sensible precautions by individuals to avoid mosquito bites reduce substantially both the risk and the transmission of malaria (Fig. 126).

5.2 Methods used to study mosquitoes

Precise identification of *Anopheles* mosquitoes is difficult, but essential for effective control of malaria. Apparent species of *Anopheles* often prove to be 'species complexes'—a cluster of separate species which look identical down to the smallest detail. Yet each species in the complex is distinct genetically, biochemically, ecologically, and in its ability to transmit malaria. Two *Anopheles* mosquitoes may look the same but belong to separate species, and be unable to interbreed. One may be a dangerous vector of malaria, the other harmless.

Avoid visiting malarious areas, especially spending the night in such parts.

Wear clothes with long sleeves and long trousers from dusk through the night till dawn.

Use insect repellents on exposed skin. A good one is diethyl toluamide (DEET) in creams, sticks and lotions.

Take prophylactic tablets as recommended.

Sensible precautions taken by individuals can reduce malaria risks and malaria transmission.

Live in screened houses. Keep the screens in good repair.

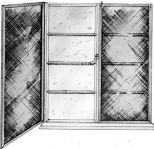

Young children in bed under a net by 7 p.m.!

Use bed-nets properly, preferably one impregnated with a residual insecticide such as permethrin.

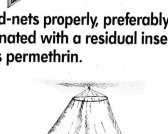

Burn mosquito coils or use vaporization mats to repel mosquitoes. Use insecticide sprays to kill mosquitoes which have entered the house.

Fortunately, modern genetic and biochemical techniques allow accurate identification of *Anopheles* species, and analysis of species complexes.

Precise identification of *Anopheles* mosquitoes is also essential for accurate planning of effective malaria control programmes. Resources will be wasted and perhaps unnecessary damage done to the environment if spraying is aimed at all mosquitoes in an area. If the important vector species are carefully identified and their behaviour analysed then action can be selective.

Crossing and interbreeding

Mosquitoes which look identical but belong to different species cannot interbreed to produce fertile offspring. The males of such 'mule' mosquitoes are sterile, with undeveloped testes. Females may not be produced at all in such a cross. Breeding experiments in the laboratory first gave evidence for the complexity of the apparent species *Anopheles gambiae*, the main vector in Africa; but this method is laborious, and has now been superseded.

Chromosome analysis

A powerful technique to identify individual species in an *Anopheles* complex is to examine the giant chromosomes in the salivary gland cells of fourth instar larvae, or the ovarian cells of females. When stained these chromosomes appear striped or banded. The banding patterns of the chromosomes of individual species form a distinctive 'fingerprint' identifying the species (Fig. 127). Different species often have segments of chromosomes inverted, revealed by the order of bands along the chromosomes.

In hybrids between two species the chromosomes fail to pair correctly during meiosis, blocking the differentiation of sperms and eggs. 'Inversion loops' occur if segments of homologous chromosomes pair with each other, and one has a segment inverted. The pairing is disordered and loops form, which are easily visible using a microscope (Fig. 128).

Salinity tests

A species complex of *Anopheles* may include species breeding in fresh water, and others breeding in brackish water. Although the fresh- and brackish-water species may look identical, they can be separated by exposing their larvae to salt concentrations of about 3 per cent. Larvae of fresh-water species will mostly die at such concentrations of salt. Mature blood-fed females captured in the area of interest are allowed to lay eggs in the laboratory. The larvae which hatch can be salt-tested, and the parent is then identified. This method gives results in a few days, and is simple and cheap. It is valuable in the identification of individual species in the important complexes *Anopheles gambiae* and *Anopheles punctulatus*, malaria vectors in Africa and Papua New Guinea respectively.

Enzyme analysis

Mosquitoes contain enzymes which will catalyse chemical reactions producing colours or precipitates. Such enzymes are easily detected. They can be separated by electrophoresis in starch or polyacrylamide gels. Molecules of each enzyme move through the gel at a distinctive

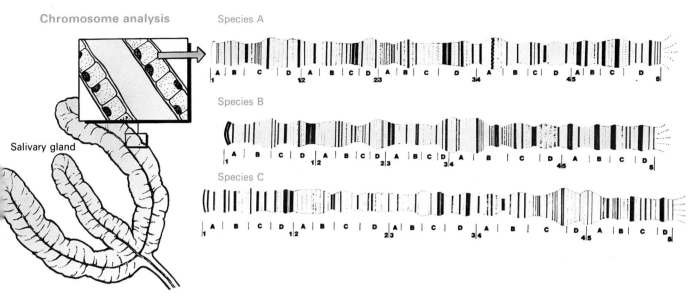

Chromosome analysis

Salivary gland

Species A

Species B

Species C

speed if the electric field, temperature, and other conditions are controlled. Reagents for the detection reaction are then applied to the gel, and colour or precipitate develops to mark the position of the enzyme. In an *Anopheles* complex electrophoresis may show that several enzymes differ in mobility among the species. The variants of a single enzyme are called isoenzymes, and the variability is called enzyme polymorphism.

An individual mosquito can be analysed in this way (Fig. 129). The insect is killed, extracted with salt solution, and the enzymes in

127. Maps of the ovarian polytene XR arms from three freshwater species of the *Anopheles gambiae* complex.

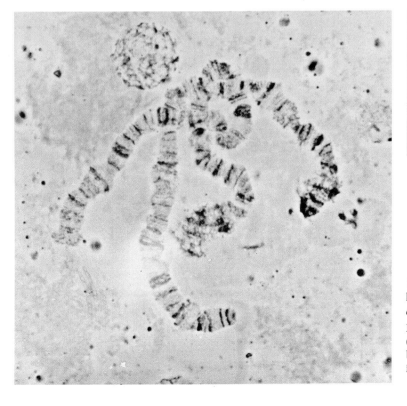

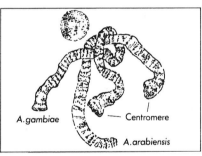

128. The X-chromosomes in an ovarian cell from a hybrid of *Anopheles gambiae* and *Anopheles arabiensis*.

Pairing of chromosomes during meiosis has failed (asynapsis) and further development of ova is blocked. The individual chromosomes are identified in the small diagram.

Dr. H. Townson/Dr. J. D. Charlwood

the extract separated and identified. By choosing the most appropriate enzyme, the resulting isoenzyme pattern is another 'fingerprint' identifying the species. Hybrids may show the isoenzymes of each parent, so this test also gives information about the extent of cross-breeding in an *Anopheles* population.

129. Isoenzymes from two closely-related *Anopheles* species.

One isoenzyme does not distinguish them, but the other identifies the top six specimens as one species and the bottom seven as the second.

Dr. H. Townson

Cuticular hydrocarbon analysis

The outer layer of mosquitoes, called the cuticle, contains waxes, fatty substances, and similar waterproofing compounds. These substances can be separated and identified by gas-liquid chromatography. They are extracted in a suitable solvent, and the solution is injected

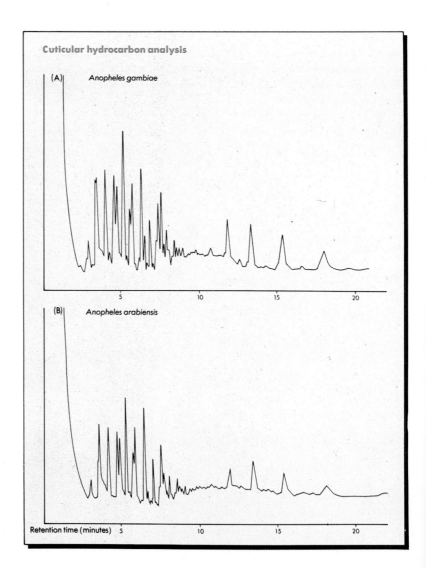

130. Gas chromatograms of crude extracts from individual female *Anopheles gambiae* complex mosquitoes. (A) *Anopheles gambiae*. (B) *Anopheles arabiensis*.

into an oven through which an inert gas is passing. Substances volatile at the oven temperature are swept through a column containing another solvent adsorbed on to a suitable solid. The volatile substances dissolve in the adsorbed solvent and re-evaporate as fresh gas passes over. The passage of different volatile substances through the column is thereby delayed to different degrees. Each substance has a characteristic retention time if other conditions are kept constant.

Gas-liquid chromatographs of crude extracts from two species of *Anopheles* are shown in Fig. 130. The pattern of cuticular constituents allows definite identification. Dried specimens can be identified in this way, including old pinned specimens from museum collections. The equipment is expensive and needs careful use and maintenance, but it is not bulky and gives rapid results.

DNA probes

The techniques of DNA technology are being developed for use in mosquito identification. DNA probes to identify mosquito species are prepared by inserting fragments of mosquito DNA into *E. coli* or another suitable bacterium (Fig. 131). The hybridization test is done

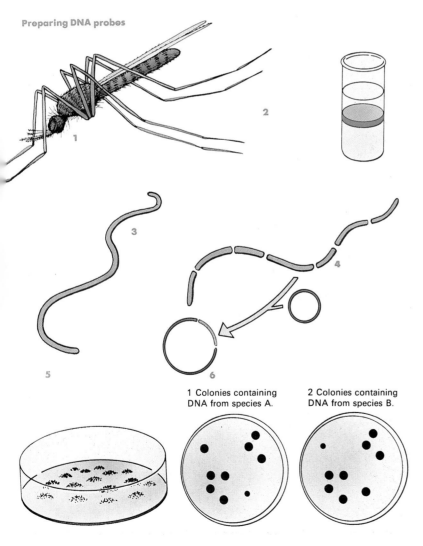

Preparing DNA probes

1 Colonies containing DNA from species A.

2 Colonies containing DNA from species B.

131. Preparing DNA probes for mosquito identification.

1. About 100 mosquitoes, apparently of the same species are ground up.
2. Genomic DNA of the mosquitoes, extracted and purified.
3. Using an enzyme the DNA is cut into pieces of a manageable size.
4. A circular plasmid DNA molecule is cut open using an enzyme, and a fragment of mosquito DNA inserted. Different fragments will be inserted in other plasmid DNA molecules. (Plasmid DNA—a small cytoplasmic DNA molecule found in bacteria.)
5. The modified plasmids are taken up by the cells of the bacterium *E.coli*. The individual bacteria are then grown into colonies. The plasmids are multiplied with the bacteria.
6. The colonies are screened with radioactive genomic DNA from each species known in the species complex. If the bacterium contains DNA derived from a mosquito species then that DNA will combine (hybridize) with radioactive DNA prepared from mosquitoes previously identified as that species, whose DNA is thus used as a probe.
1. Colonies containing DNA from species A.
2. Colonies containing DNA from species B.

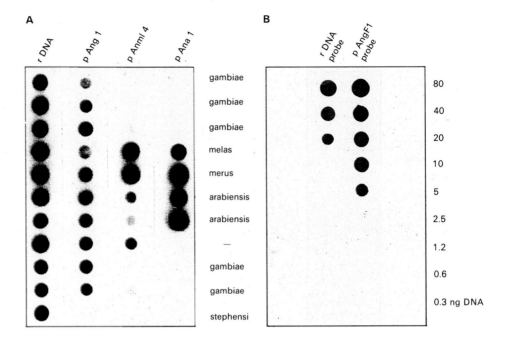

132. The 'dot-blot' procedure for mosquito identification.

A. Dot-blot hybridization of four DNA probes to different species of the *Anopheles gambiae* complex. pAna 1 will differentiate *Anopheles arabiensis* from *Anopheles gambiae* but not from *Anopheles melas* or *Anopheles merus*. pAnml 4 will differentiate *Anopheles melas* and *Anopheles merus* from both *Anopheles gambiae* and *Anopheles arabiensis*. Hence together these two probes permit the separation of *Anopheles gambiae* and *Anopheles arabiensis* from each other and from the two salt-water species, *Anopheles melas* and *Anopheles merus*. pAng 1 rDNA differentiates *Anopheles gambiae* from other *Anopheles* species. It is a ribosomal DNA probe used to titrate the intensity of the dot blot.

B. Titration of *Anopheles arabiensis* DNA against 2 molecular probes. pAngF 1 is a complementary DNA probe; rDNA is a segment of mosquito ribosomal DNA. *Dr. H. Townson/Dr. J. D. Charlwood*

by blotting the colonies on to nitrocellulose paper, to which solutions of radioactive probe DNA are applied. One variety of this test is the 'dot-blot' procedure (Fig. 132). Tests are now so sensitive that a species can be identified from the spermatozoa injected into a female by one male mosquito at copulation. Most tests require radioactive DNA markers, which restricts their potential application. Colour reactions are now under intensive study, so that DNA techniques can be used in simple test-kits for large-scale field use.

5.3   Mosquito control in Papua New Guinea

This section is a case-study of malaria control in Papua New Guinea. It illustrates the techniques and value of careful mosquito identification, and the difficulty of predicting the results of casually planned insecticide application.

133. Map of the South West Pacific Region.

Introduction

Malaria is widespread in the south-west Pacific (Fig. 133). It is a major public health problem in Papua New Guinea. Each of the children in the picture (Fig. 134) had malaria parasites in the blood. The predominant malaria parasite is *Plasmodium falciparum*, but *Plasmodium vivax* and *Plasmodium malariae* are common: up to a quarter of infections in some areas. *Plasmodium ovale* is rare.

In the 1960s there were hopes that malaria might be eliminated by the indoor spraying of houses with residual insecticides. By the 1980s it had become abundantly clear that this approach would not succeed. Indeed in coastal areas of Papua New Guinea the incidence of malaria was as high as it had been before spraying started. The reasons for this disappointment are many and complicated, but the behaviour of the mosquito vectors is important. Development of resistance to chloroquine in *Plasmodium falciparum* has further increased the difficulties of malaria control in the region.

Main vectors

Mosquitoes of the *Anopheles punctulatus* group (Fig. 135) are the most important vectors of malaria in Papua New Guinea. The group contains a number of species which are similar morphologically but biologically diverse: individual species varying in ecology and behaviour. Such differences affect the capacity of a species to transmit malaria, and the ease of its control.

There are three species complexes, each identified by the scale patterns on the proboscis (Fig. 136):

- *Anopheles punctulatus*;
- *Anopheles koliensis*; and
- *Anopheles farauti*.

However, these complexes contain at least six biological species, and undoubtedly others await recognition. Thus the species complex *Anopheles farauti* appears to consist of at least four distinct species, which look identical, but can be recognized by the banding patterns of their polytene chromosomes. They are called *Anopheles farauti* 1, 2, 3, and 4 respectively.

In Papua New Guinea *Anopheles punctulatus*, *Anopheles koliensis*, and *Anopheles farauti 1* occur. *Anopheles koliensis* is at least two distinct species, living in the same habitat but not interbreeding. The differences in biology between the species which occur in Papua New Guinea have· an important bearing on malaria transmission and control.

134. Children in Papua New Guinea with the coconut harvest. Each one has malaria parasites in the blood for most of the time.

Dr. J. D. Charlwood

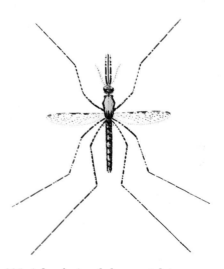

135. A female *Anopheles punctulatus*.

Cambridge Animal and Public Health, Ltd.

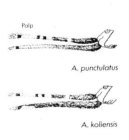

Palp

A. punctulatus

A. koliensis

Proboscis

A. farauti

136. Palp and proboscis markings in three of the *Anopheles* species complexes found in Papua New Guinea: *Anopheles punctulatus*, *Anopheles koliensis*, and *Anopheles farauti*.

137. A breeding site for *Anopheles farauti*: fresh or brackish water, coasts and edges of streams.

Dr. J. D. Charlwood

Breeding site preference

The type of water selected by each species for breeding can determine the ease of control. It may affect the proximity of breeding site to village, the interval between successive blood meals, and any seasonal variation in malaria transmission.

Breeding site preference also influences the effect of rain on mosquito numbers. For example, *Anopheles punctulatus* and *Anopheles koliensis* increase during the rainy season, but *Anopheles farauti 1* is reduced with rainfall, because its larvae are flushed from their brackish-water breeding sites (Fig. 137). This in turn influences malaria transmission. In an inland village (Fig. 138), *Anopheles punctulatus* causes peak transmission of malaria in the middle of the rains; but in coastal villages peak transmission is caused by *Anopheles farauti 1* as the rains end.

Feeding behaviour

The feeding habits of a species of *Anopheles* determine whether the species will transmit malaria effectively or not. An efficient vector species is anthropophilic: it will usually feed on man in preference to domestic or other animals. A species which is zoophilic feeds on animals by choice, but will feed on man if it must. It will not be an efficient vector.

In Papua New Guinea *Anopheles punctulatus* is anthropophilic and an important vector. *Anopheles farauti 1* is an indiscriminate feeder, often switching from biting humans to biting domestic animals, especially pigs. It is a vector of malaria, but not such an important one as *Anopheles punctulatus*. Most coastal villages have many pigs and other animals, which distract *Anopheles farauti 1* from people (Fig. 139).

Feeding habits

Anopheles farauti 1, *Anopheles punctulatus*, and *Anopheles koliensis* differ in feeding habits, and each species has a different feeding time, indoors or out. *Anopheles farauti 1* is most active in the evening, biting before midnight, and feeding indoors and out. Consequently people are at greatest risk from *Anopheles farauti 1* when they gather socially in the evening, before going to sleep (Fig. 140).

By contrast, *Anopheles punctulatus* and *Anopheles koliensis* feed mostly after midnight, indoors, biting sleeping people. Under such circumstances bed-nets may provide protection against malaria.

Resting sites

Many malaria vectors rest indoors on walls and ceilings before and after feeding. This habit is important for malaria control, because such resting sites are good targets for persistent insecticides. A mosquito species which rests is easier to control than another which does not.

House construction affects the entry of mosquitoes, and their resting habits. In Papua New Guinea *Anopheles punctulatus* rests indoors after feeding until it is ready to lay its eggs, but *Anopheles farauti 1* may not rest indoors after feeding, and when it does, it leaves with the following dawn. The open structure of houses in coastal areas facilitates entry and exit by this species (Fig. 141).

138. An inland village infested by *Anopheles punctulatus*. Peak transmission of malaria is in the middle of the rains.

Dr. J. D. Charlwood/Dr. P. Graves

139. In this coastal village *Anopheles farauti 1* has a wide range of alternative hosts. Domestic pigs are the most attractive to the mosquito because they have large areas of bare skin.

Dr. J. D. Charlwood

140. A social evening in Papua New Guinea. This group will very likely be bitten by *Anopheles farauti 1*.   *Dr. J. D. Charlwood*

141. Dawn in Papua New Guinea. The open structure of housing in coastal villages facilitates the entry and exit of *Anopheles farauti 1*.

Dr. J. D. Charlwood

House spraying

The traditional method to control malaria vectors is to spray insecticides on to surfaces inside houses. The results of such spraying can be greatly influenced by mosquito behaviour. For example,in the Solomon Islands indoor spraying with DDT resulted in the disappearance of *Anopheles punctulatus* and *Anopheles koliensis* after two spray rounds, whereas *Anopheles farauti 1* persisted, and changed its behaviour, living more outdoors and feeding more on man (Fig. 142). Similar changes were seen in Western Irian Jaya. Such changes in behaviour are difficult to explain. They may be the effect of DDT selecting a particular behaviour pattern, or the consequence of subtle genetic differentiation of *Anopheles farauti* into sub-groups or species.

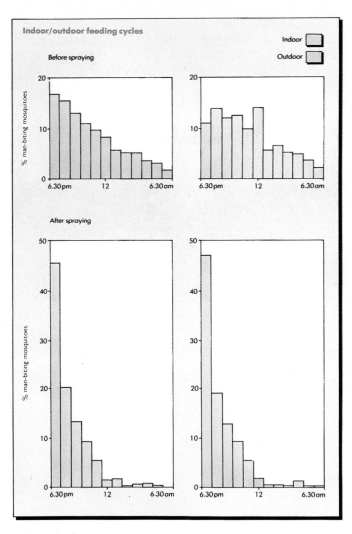

142. Outdoor and indoor feeding by *Anopheles farauti* before and after DDT spraying of houses.

Impregnated bed-nets

Bed-nets have been used for many years, but there are several practical problems in their use. Damaged bed-nets, or nets not properly tucked in, are ineffective, and it isn't easy to tuck in your own. To wake in the morning badly bitten and find blood-engorged mosquitoes resting inside the net is a poor incentive to continue using it.

143. After drying treated nets are ready to give another six months' protection.

Dr. J. D. Charlwood

A possible solution to these problems is to impregnate nets with Permethrin, a synthetic insecticide. Nets with large holes, or ones which have not been properly tucked in, may then give good protection. Impregnated nets were effective and acceptable in trials in Papua New Guinea, and in other tropical areas. The community can become involved in making and treating nets. Families bring their nets to treatment centres for impregnation with insecticide. After treatment the nets are partially dried by leaving them flat on the ground. Drying is completed by hanging the nets up (Fig. 143).

A properly impregnated net provides protection for at least six months, and holes or careless use matter much less than they do with untreated nets. Treatment costs over a four-year period are less than those of routine house spraying with DDT. Impregnated nets provide relief from other night-biting insects, and are readily accepted in the community (Fig. 144).

Trial 1	Total feed (6 nights)
Room 1 (without nets)	1394
Room 2 (with nets)	273

Trial 2	Total feed (3 nights)
Room 1 (with nets)	109
Room 2 (with nets)	149

144. Numbers of mosquitoes collected from the walls of two bedrooms in Budip village, in Papua New Guinea.

5.4   Malaria eradication in County Rong, Sichuan

This section is another case study, this time of a malaria eradication project which is close to success.

The hilly County Rong is in Sichuan Province, in the south western part of The People's Republic of China. It has an area of 1952 square kilometres and a population of 820 000. There are 7 districts, 70 townships, 587 villages, and 5042 production groups. Some 6000 personnel at various levels work full-time or part-time in the antimalarial programme.

The season of malaria transmission lasts from May to September. So far, only *Plasmodium vivax* has been found. *Anopheles sinensis* is

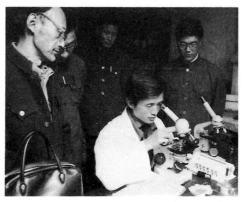

145. Training a health-worker to recognize malaria parasites in a blood-smear. *Dr. Zhao-Xi Zhou*

146. A health-worker watches a malaria carrier, ensuring that the treatment is taken.

Dr. Zhao-Xi Zhou

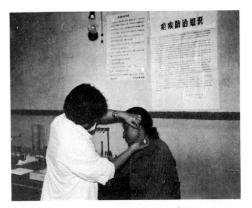

147. Taking a blood-smear from a fever case at one of the 70 microscopy units. *Dr. Zhao-Xi Zhou*

148. Educational poster showing areas of high prevalence of malaria in County Rong.

Dr. Zhao-Xi Zhou

the main malaria vector. *Anopheles lesteri anthropophagus* is also widely distributed, and *Anopheles minimus* occurs in a limited area.

Malaria was prevalent in the county, and an outbreak in 1962 caused about 180 000 cases. After this surveillance was increased. Cases with symptoms of malaria during the previous two years were identified, investigated, and treated if necessary (Figs. 145–148). Preventive mass treatment was carried out in villages and production groups reporting an incidence of malaria greater than 5 per cent in one year, and in those located near the border of the county.

Cycloguanil was the usual antimalarial drug available before 1965, but since then chloroquine, primaquine, and pyrimethamine have been used widely. For treatment of malaria choloroquine is given: 600 mg (base) as a loading dose and 300 mg on each of the following two days for adults. At the same time a course of primaquine, 22.5 mg daily for eight days is given to eradicate hepatic forms. For prophylaxis pyrimethamine 50 mg is given every two weeks. Treatment is directly supervised.

The number who had taken antimalarial drugs was estimated at 1 390 000 during 1962–1974. Control of the outbreak was achieved within a few years.

In 1975 malaria recurred, with an incidence of 1557, during the social dislocations of the Cultural Revolution. Therefore County Rong was designated a demonstration area, with the emphasis shifted from malaria control to its eventual eradication.

Surveillance now covers all the population. During 1978–1985 564 387 blood-smears of fever cases were made. Among them 155 malaria cases were identified and treated. 173 174 persons travelling between endemic areas and the county were given drugs: a single dose of chloroquine for a short visit, or one combined with primaquine for longer residence.

All five neighbour counties were invited to join with County Rong in a common antimalarial programme, with good control results. The morbidity of malaria in County Rong has now been less than 0.01 per cent for eight successive years.

5.5 Imported malaria in Europe

Imported malaria refers to cases acquiring the disease outside a specified country or region. In Britain all malaria is now imported. All travellers from Britain to areas where malaria transmission occurs need to follow advice on drug prophylaxis and avoiding mosquito bites. Probably fewer than half do. The statistics on imported malaria are collected by the Malaria Reference Laboratory in the London School of Hygiene and Tropical Medicine.

'Airport malaria' is now an occupational risk of airport workers, particularly baggage handlers and maintenance crew, who may be bitten by mosquitoes which entered the aircraft in a malarious region.

Increasing case numbers

The number of cases of malaria imported into Britain has increased from 101 in 1970 to 2212 in 1985. Factors contributing to this rise in incidence include the rapid expansion of air travel, the resurgence of malaria, and the spread of chloroquine resistance in *Plasmodium.*

Most cases of imported malaria occur in travellers from Asia and Africa (Fig. 149). Immigrants from Asia living in Britain account for more than half of all cases of imported malaria. Nigerians and Ghanaians are the visitors most likely to develop malaria while in Britain.

Advice for travellers

Those travellers most at risk of contracting malaria are least likely to seek advice about protection. Travellers frequently 'shop around' for information, and become confused by the different regimes which may be recommended. Travellers often cannot accurately recall advice given. If more than one prophylactic regime is recommended then understanding and recall are made worse. Less than half the travellers surveyed complied fully with advice. Non-compliance occurs mainly on return (Fig. 150).

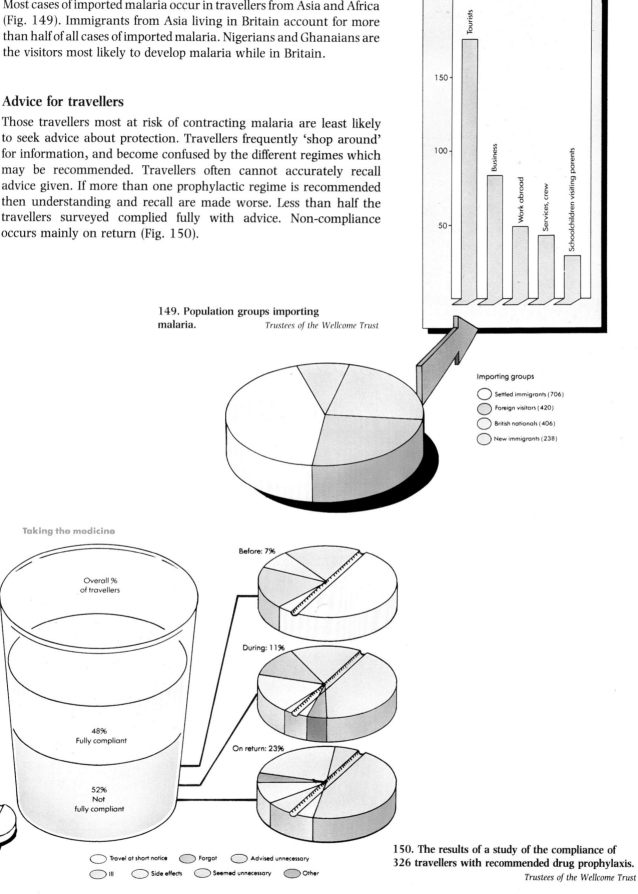

149. Population groups importing malaria.   *Trustees of the Wellcome Trust*

150. The results of a study of the compliance of 326 travellers with recommended drug prophylaxis.
Trustees of the Wellcome Trust

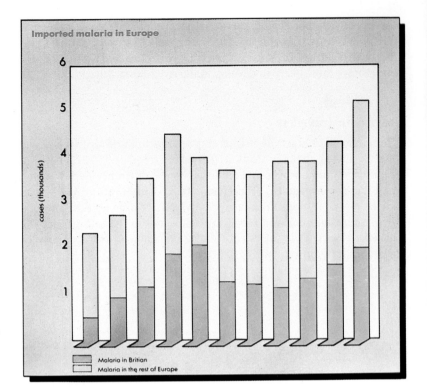

151. Imported malaria in Europe.

Numbers of cases reported to the WHO, 1974 to 1985. *Trustees of the Wellcome Trust*

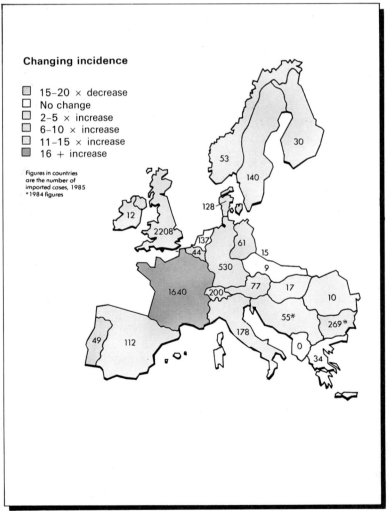

Imported malaria in Europe (Fig. 151)

Malaria imported into Britain accounts for 40 per cent of all European cases reported to the World Health Organization. In Europe the epidemiological characteristics of imported malaria remain fairly constant. They reflect traditional patterns of travel and commerce with the countries concerned, and are often still determined by previous colonial relationships.

1. Belgium

Five cases of 'airport' malaria were reported in 1986, of which one died. Over half Belgium's imported cases originate in Zaire.

2. Bulgaria

Three-quarters of cases reported were in new immigrants from Ethiopia, Angola, and Niger.

3. France

Fifty-one cases were reported in 1984, but numbers had soared to 1640 in 1985. Seventy-eight per cent of infections originate in Africa.

4. West Germany

Seventeen deaths were reported in 1985, ten of which were in tourists who had visited Africa.

5. Greece

Seventy per cent of cases reported occur in sailors. Eight cases were due to blood transfusions containing malaria parasites.

6. Portugal

Ninety per cent of reported cases are caused by *Plasmodium falciparum*, half of which originate in Angola.

At present the incidence of imported malaria, the information collected and reported on each individual, the coverage of case reporting, and the advice given on malaria prophylaxis all vary greatly between countries. Comparisons are therefore difficult.

A wide variety of prophylactic drugs and regimes are recommended to Europeans. Only two of eleven countries surveyed had a national policy on malaria drug prophylaxis, and twenty-one different prophylactic regimes were recorded. The coverage and reliability of reporting vary, both between countries and also between ethnic and other sub-groups within countries. Standardization of the collection of data is needed, to permit accurate analysis of the risks, management, and prevention of imported malaria.

5.6 Prospects

Malaria eradication is technically feasible but so far unobtainable. New drugs, vaccines, and methods to combat mosquitoes may each facilitate malaria control, but the real problem is to relieve the poverty exposing so many tropical populations to the blight of malaria and other infectious diseases.

We must not expect to conquer malaria, a disease rooted in the physical environment of tropical areas and in their unsatisfactory socio-economic conditions, by a miraculous vaccine, drug or insecticide. Each of such scientific tools will be of great value but their proper use will depend on other factors closely related to human ecology in its broadest sense and especially on the national will to fight the disease and on the international determination to co-operate with the developing countries in this endeavour.

The difficult problems of malaria eradication or of its control in relation to the development of the Third World exemplify in a small way much of our present predicament. Higher standards of health and education depend on more wealth to provide for them but health and education are equally necessary to increase economic production. And yet a balanced economic development combined with increased food production per head cannot advance where the population pressure is too high. On the other hand any excessive increase of population will never be solved by neglecting public health and medical care.

(L. J. BRUCE-CHWATT 1979).

Appendix
Severe malaria: a medical emergency

Plasmodium falciparum is poorly adapted to its human host, causing severe or lethal illness in non-immune patients. Throughout history it has plagued armies, merchants, missionaries, and other travellers. It remains a serious threat to susceptible people in many tropical areas, especially to children and to pregnant women.

The illness has many names, including pernicious malaria, cerebral malaria, and malignant tertian fever. Blackwater fever is an unusual variant.

Severe *falciparum* malaria is the name now preferred. Severe malaria is a clinical emergency, where prompt diagnosis and treatment may be life-saving.

Severe *falciparum* malaria is the cause of almost all malaria deaths, probably between 1 and 2 million each year.

A.1   Sequestration in severe malaria

The development of *Plasmodium falciparum* in the blood differs from the development of other human species of malaria.

Erythrocytes containing parasites adhere to the endothelial cells lining the smallest veins of internal organs, a process called sequestration. In these immobilized red cells the parasites complete their growth and segmentation into merozoites. The erythrocytes finally burst, freeing the merozoites into the circulating blood, where they can infect new red cells and repeat the cycle.

Knobs develop on the surface of erythrocytes containing *Plasmodium falciparum*. The knobs contain a histidine-rich protein secreted by the parasite. This interacts with a component of the endothelial cell membrane, causing the red cell to adhere.

Adherence and sequestration of infected red cells is intense in the brain, and causes a measurable reduction in brain blood flow. This may explain why confusion, coma, and fits are common in this disease.

The intestines are also badly affected by sequestration, and diarrhoea is a frequent symptom. In pregnant women the placenta is often the worst affected organ, and abortion is very common.

Sequestration is much reduced in patients lacking the spleen, but the reason for this is unknown. Sequestration is also suppressed in people indigenous to endemic malarious areas. Again the reason is unknown, but it appears to be in part an immune process, contributing to the tolerance of the parasite by individuals in such populations.

Abnormal immune responses may contribute to the disease process in severe malaria. Tumour Necrosis Factor (TNF) is a peptide released from macrophages during immune reactions, and causes clinical features comparable to malaria if infused. Overproduction of TNF has been reported in malaria, but the significance of such observations is still obscure.

A.1   Sequestration in severe malaria.

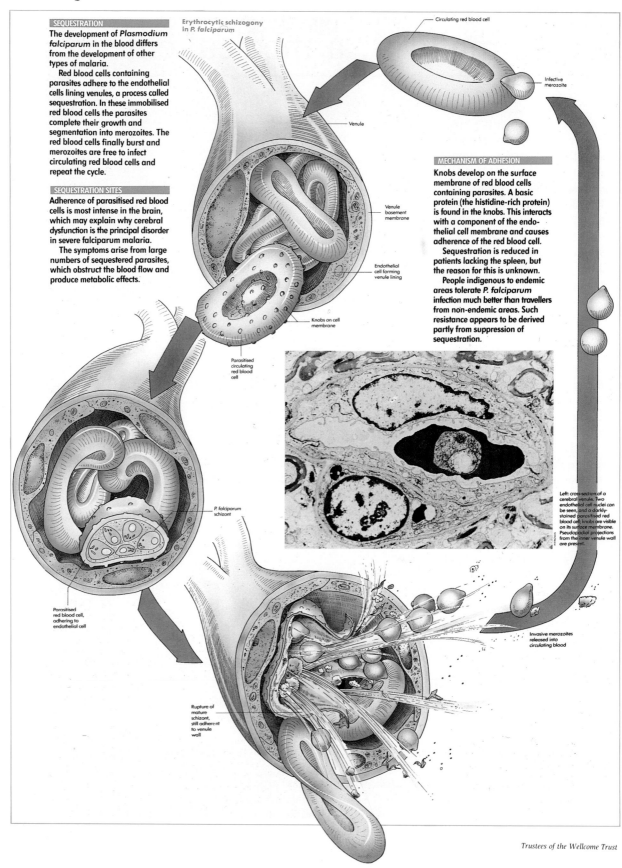

SEQUESTRATION

The development of *Plasmodium falciparum* in the blood differs from the development of other types of malaria.

Red blood cells containing parasites adhere to the endothelial cells lining venules, a process called sequestration. In these immobilised red blood cells the parasites complete their growth and segmentation into merozoites. The red blood cells finally burst and merozoites are free to infect circulating red blood cells and repeat the cycle.

SEQUESTRATION SITES

Adherence of parasitised red blood cells is most intense in the brain, which may explain why cerebral dysfunction is the principal disorder in severe falciparum malaria.

The symptoms arise from large numbers of sequestered parasites, which obstruct the blood flow and produce metabolic effects.

Erythrocytic schizogony in *P. falciparum*

Circulating red blood cell

Infective merozoite

Venule

Venule basement membrane

Endothelial cell forming venule lining

Knobs on cell membrane

Parasitised circulating red blood cell

MECHANISM OF ADHESION

Knobs develop on the surface membrane of red blood cells containing parasites. A basic protein (the histidine-rich protein) is found in the knobs. This interacts with a component of the endothelial cell membrane and causes adherence of the red blood cell.

Sequestration is reduced in patients lacking the spleen, but the reason for this is unknown.

People indigenous to endemic areas tolerate *P. falciparum* infection much better than travellers from non-endemic areas. Such resistance appears to be derived partly from suppression of sequestration.

P. falciparum schizont

Parasitised red blood cell, adhering to endothelial cell

Rupture of mature schizont, still adherent to venule wall

Invasive merozoites released into circulating blood

Left: cross-section of a cerebral venule. Two endothelial cell nuclei can be seen, and a darkly-stained parasitised red blood cell; knobs are visible on its surface membrane. Pseudopodial projections from the inner venule wall are present.

A.2 Clinical features of severe malaria

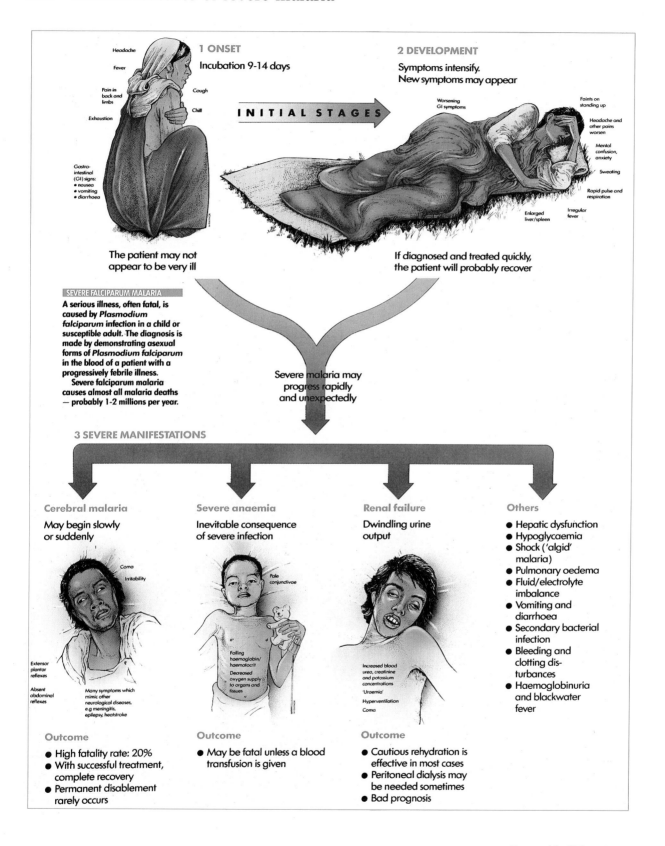

1 ONSET

Incubation 9-14 days

Headache
Fever
Pain in back and limbs
Cough
Chill
Exhaustion
Gastro-intestinal (GI) signs:
● nausea
● vomiting
● diarrhoea

The patient may not appear to be very ill

INITIAL STAGES

2 DEVELOPMENT

Symptoms intensify.
New symptoms may appear

Worsening GI symptoms
Faints on standing up
Headache and other pains worsen
Mental confusion, anxiety
Sweating
Rapid pulse and respiration
Enlarged liver/spleen
Irregular fever

If diagnosed and treated quickly, the patient will probably recover

SEVERE FALCIPARUM MALARIA

A serious illness, often fatal, is caused by *Plasmodium falciparum* infection in a child or susceptible adult. The diagnosis is made by demonstrating asexual forms of *Plasmodium falciparum* in the blood of a patient with a progressively febrile illness.

Severe falciparum malaria causes almost all malaria deaths — probably 1-2 millions per year.

Severe malaria may progress rapidly and unexpectedly

3 SEVERE MANIFESTATIONS

Cerebral malaria

May begin slowly or suddenly

Coma
Irritability
Extensor plantar reflexes
Absent abdominal reflexes
Many symptoms which mimic other neurological diseases, e.g meningitis, epilepsy, heatstroke

Outcome
● High fatality rate: 20%
● With successful treatment, complete recovery
● Permanent disablement rarely occurs

Severe anaemia

Inevitable consequence of severe infection

Pale conjunctivae
Falling haemoglobin/haematocrit
Decreased oxygen supply to organs and tissues

Outcome
● May be fatal unless a blood transfusion is given

Renal failure

Dwindling urine output

Increased blood urea, creatinine and potassium concentrations
'Uraemia'
Hyperventilation
Coma

Outcome
● Cautious rehydration is effective in most cases
● Peritoneal dialysis may be needed sometimes
● Bad prognosis

Others

● Hepatic dysfunction
● Hypoglycaemia
● Shock ('algid' malaria)
● Pulmonary oedema
● Fluid/electrolyte imbalance
● Vomiting and diarrhoea
● Secondary bacterial infection
● Bleeding and clotting disturbances
● Haemoglobinuria and blackwater fever

A.3   Intensive care of severe malaria

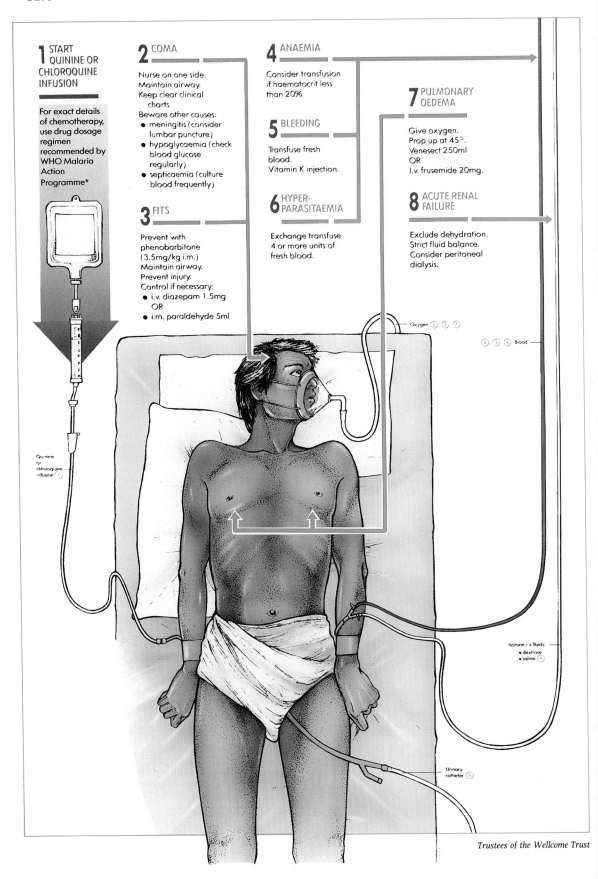

1 START QUININE OR CHLOROQUINE INFUSION

For exact details of chemotherapy, use drug dosage regimen recommended by WHO Malaria Action Programme*

2 COMA

Nurse on one side.
Maintain airway.
Keep clear clinical charts
Beware other causes:
● meningitis (consider lumbar puncture)
● hypoglycaemia (check blood glucose regularly)
● septicaemia (culture blood frequently)

3 FITS

Prevent with phenobarbitone (3.5mg/kg i.m.)
Maintain airway.
Prevent injury.
Control if necessary:
● i.v. diazepam 1.5mg
   OR
● i.m. paraldehyde 5ml

4 ANAEMIA

Consider transfusion if haematocrit less than 20%

5 BLEEDING

Transfuse fresh blood.
Vitamin K injection.

6 HYPER-PARASITAEMIA

Exchange transfuse 4 or more units of fresh blood.

7 PULMONARY OEDEMA

Give oxygen.
Prop up at 45°.
Venesect 250ml
OR
I.v. frusemide 20mg.

8 ACUTE RENAL FAILURE

Exclude dehydration.
Strict fluid balance.
Consider peritoneal dialysis.

Oxygen ② ③ ⑦

④ ⑤ ⑥ Blood

Quinine or chloroquine infusion ①

Isotonic i.v. fluids:
● dextrose
● saline ⑧

Urinary catheter ⑧

A.4 Severe *falciparum* malaria: antimalarial chemotherapy in adults and children

Facilities intended to provide maximum possible health care

Chloroquine-sensitive	Chloroquine-resistant or sensitivity unknown[a]
1. *Chloroquine*: 10 mg *base*/kg in isotonic fluid by constant rate i.v. infusion over 8 hours, followed by 15 mg/kg over 24 hours.	**1.** *Quinine*: 7 mg dihydrochloride *salt*/kg (loading dose)[b] i.v. by infusion pump over 30 minutes followed immediately by 10 mg/kg diluted in 10 ml/kg isotonic fluid by i.v. infusion over 4 hours, repeated 8 hourly (maintenance dose) until the patient can swallow, then quinine tablets approx. 10 mg *salt*/kg 8 hourly to complete 7 days' treatment.
OR **2.** *Chloroquine*: 5 mg *base*/kg in isotonic fluid by constant rate i.v. infusion over 6 hours to a total dose of 25 mg *base*/kg over 30 hours.	OR **2.** *Quinine*: 20 mg *salt*/kg (loading dose)[b] by infusion over 4 hours, then 10 mg/kg over 4 hours, 8 hourly until patient can swallow, then quinine tablets to complete 7 days' treatment.
OR **3.** *Quinine*: (see right-hand column)	OR **3.** *Quinidine*: 10 mg gluconate *salt*/kg (loading dose)[b] by infusion over 1–2 hours, followed by 0.02 mg/kg/min by infusion pump for 72 hours or until the patient can swallow, then quinine tablets to complete 7 days of treatment.
	OR **4.** *Quinidine*: 15 mg gluconate/kg (loading dose) by i.v. infusion over 4 hours, then 7.5 mg/kg over 4 hours, 8 hourly until patient can swallow, then quinine tablets to complete 7 days' treatment.

[a] In areas with a significant degree of quinine-resistance (e.g. Thailand) add an oral course of tetracycline 250 mg qds for 7 days except for children under 8 years and pregnant women.
In patients requiring more than 48 hours of parenteral therapy halve the quinine or quinidine maintenance dose to 5 mg/kg.
[b] Loading dose should not be used if patient received quinine, quinidine, or mefloquine within the preceding 12–24 hours.

Facilities intended to provide intermediate health care

Chloroquine-sensitive	Chloroquine-resistant or sensitivity unknown[a]
1. *Chloroquine*: 3.5 mg *base*/kg six hourly given intramuscularly (anterior thigh) or subcutaneously. Change to oral as soon as patient can swallow tablets to complete total dose 25 mg *base*/kg.	**1.** *Quinine*: 20 mg dihydrochloride/kg (loading dose)[b] given intramuscularly (divided sites, anterior thigh), then 10 mg/kg 8 hourly until patient can swallow, then quinine tablets approx. 10 mg *salt*/kg 8 hourly to complete 7 days' tretment.
2. *Quinine*: (see right-hand column)	

[a] In areas with a significant degree of quinine resistance (e.g. Thailand) add tetracycline (see above). In patients requiring more than 48 hours of parenteral therapy halve the quinine maintenance dose to 5 mg/kg.
[b] Loading dose should not be used if patient received quinine, quinidine or mefloquine within the preceding 12–24 hours.

Facilities intended to provide minimum health care (periphery/community)

Chloroquine-sensitive	Chloroquine-resistant or sensitivity unknown
1. *Chloroquine*: tablets/syrup by mouth[a]	**1.** *Quinine*: tablets by mouth 10 mg *salt*/kg[a]
then refer patient to higher level for parenteral treatment	*then* refer patient to higher level for parenteral treatment
OR continue 5 mg *base*/kg at 6, 24, and 48 hours later	OR continue 10 mg *salt*/kg 8 hourly to complete 7 days' treatment[b]
? suppositories	
2. *Quinine, mefloquine,* or *sulfadoxine/pyrimethamine* (see right-hand column)	**2.** *Mefloquine*: tablets by mouth[a], 15 mg *base*/kg single dose
	3. *Sulfadoxine-pyrimethamine*[c,d]: tablets by mouth[a], sulfadoxine 25 mg/kg pyrimethamine 1.25 mg/kg single dose.

[a] Unless patient cannot take medication reliably by mouth.
[b] In areas with a significant degree of quinine resistance (e.g. Thailand) add an oral course of tetracycline 250 mg qds for 7 days except for children under 8 years and pregnant women.
[c] Not in pregnant or lactating women, unless there is no alternative.
[d] In some countries sulfalene-pyrimethamine is used instead.

Suggestions for further reading

Books

Bruce-Chwatt, L.J. (1985).
Essential malariology.
Heinemann Medical Books, London.
[Concise and indispensable.]

Bruce-Chwatt, L.J. and de Zulueta, J. (1980).
The rise and fall of malaria in Europe.
Oxford University Press.

Garnham, P.C.C. (1966).
Malaria parasites and other Haemosporidia.
Blackwell, Oxford.
[A classic on the biology of malaria parasites.]

Grammiccia, G. (1988).
The life of Charles Ledger, (1818–1905).
The Macmillan Press, Basingstoke.
[The extraordinary history of the man who gave quinine
   to the world.]

Harrison, G. (1978).
Mosquitoes, malaria and man: a history of the hostilities since 1880.
John Murray, London.
[An excellent read.]

Wernsdorfer W.H. and MacGregor, Sir Ian (1988).
Malaria (2 volumes)
Churchill Livingstone.
[Comprehensive, modern, advanced textbook.]

Reviews

Several authors.
Malaria.
British Medical Bulletin (1982), **vol.38**, no.2.

Bruce-Chwatt, L. J. (1979).
Man against malaria: conquest or defeat.
*Transactions of the Royal Society of Tropical
   Medicine and Hygiene*, 73, 605–617.

Clark, I. A. (1987).
Cell-mediated immunity in protection and
   pathology of malaria.
Parasitology Today, **3**, 300–305.

Cook, G. C. (1988).
Prevention and treatment of malaria.
Lancet, **1**, 32–36.

Kemp, D. J., Coppel R. L., and Anders, R. F.
   (1987)
Repetitive proteins and genes of malaria.
Annual Review of Microbiology, **41**, 181–208.

Malaria and the red cell, Ciba Foundation
Symposium 94. (1983).
Pitman, London.

World Health Organization (1986).
Severe and complicated Malaria.
*Transactions of the Royal Society of Tropical
   Medicine and Hygiene*, **80** (supplement),
   1–50.

World Health Organization (1990).
Severe and complicated malaria.
*Transactions of the Royal Society of Tropical
   Medicine and Hygiene*, **84** (supplement 2),
   1–65.

Acknowledgements

Academic Press Ltd. Fig. 107; Aikawa, Prof. M. Fig. 94; *Annals of Tropical Medicine and Parasitology* Fig. 127; Armed Forces Institute of Pathology, Washington D.C. Fig. 43; Athlone Press Fig. 7; Bannister, Dr L. H. Figs. 35, 95, 98, 99; Bruce-Chwatt, Prof. L. J. Figs. 63, 111, 112, 117, 118, 120; Cambridge Animal & Public Health Ltd. (CAMCO). Fig. 135; Charlwood, Dr J. D. Figs. 48, 134, 137, 138, 139, 140, 141, 143; Cochrane, Dr A. H. Fig. 107; Éditions André Sauret Fig. 5; Edm. & Et Sergent and L. Parrot, *La decouverte de Laveran* (1880) Figs. 18, 19; Food & Agriculture Organization Fig. 49; Gillett, Prof. J. D. Fig. 41; Graves, Dr P. Fig. 138; Greenberg Publishers, *The Conquest of Malaria*, Jaramillo-Arango, J. (1950) Fig. 11; *Illustrations of the Nueva Quinologia of Pavon*, W. Fitch F.L.S., J. Eliot Howard, F.L.S., Lovell Reeve and Co. (1862) Fig. 10; Imperial War Museum Fig. 9; Kemp, Dr. D. J. Figs. 106, 110; Krotoski, Dr W. A. Fig. 91; London Scientific Films Ltd. Fig. 42; Manson-Bahr, P. H. *Int. Rev. Trop. Med.* 2 Fig. 22; Marsden, Prof. P. Figs. 53, 57; Meddia, Royal Tropical Institute, The Netherlands Figs. 64, 114; Meis, Dr J. F. G. M. Figs. 34, 71, 88, 89, 90, 92; Mosquito Research and Control Unit, The Cayman Islands Figs. 44, 122; Peters, Prof. W. Figs. 60, 61; Ross Institute, London School of Hygiene and Tropical Medicine Fig. 16; Ross, R. (1902), *Researches on Malaria* Fig. 20; Royal Air Force Institute of Pathology and Tropical Medicine, Halton Fig. 3; Royal Pharmaceutical Society of Great Britain Figs. 14, 15; Royal Society of Tropical Medicine, Table VI from the supplement 'Severe and Complicated Malaria' *Transactions of the Royal Society of Tropical Medicine* Vol. 80, 1986 Table A.4 courtesy of the WHO; Sergeant, Prof. G. R. Fig. 56; Sinden, Prof. R. E. Figs. 29, 31, 32, 73, 77, 78, 79, 81, 83, 84, 85, 95, 101, 104; TALC Fig. 50; Townson, Dr H. Figs. 128, 129, 132; University of California Press, *The Mosquitoes of the South Pacific*, Belkin, J. N. (1962) Fig. 136; Warrell, Prof. D. A. Figs. 4, 23; WHO Figs. 28, 37, 38, 39, 40, 119; Zhao-Xi Zhou, Dr Figs. 145, 146, 147, 148.

Cover illustrations: *Top* Aerial spraying of insecticide in the West Indies. Mosquito Research and Control Unit, Cayman Islands; *Centre* Severe falciparum malaria in Thailand. Prof. D. A. Warrell; *Bottom* Tropical towns do not escape malaria. TALC.

Frontispiece: Shows an oocyst on the gut wall of a mosquito. Dr J. D. Charlwood, Dr Patricia Graves.

Index